Lecture Notes in Computer Science 13166

More information about this subseries at https://link.springer.com/bookseries/7412

Marc Aubreville · David Zimmerer ·
Mattias Heinrich (Eds.)

Biomedical Image Registration, Domain Generalisation and Out-of-Distribution Analysis

MICCAI 2021 Challenges: MIDOG 2021,
MOOD 2021, and Learn2Reg 2021
Held in Conjunction with MICCAI 2021
Strasbourg, France, September 27 – October 1, 2021
Proceedings

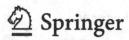 Springer

Editors
Marc Aubreville 🆔
Technische Hochschule Ingolstadt
Ingolstadt, Germany

David Zimmerer 🆔
German Cancer Research Center (DKFZ)
Heidelberg, Germany

Mattias Heinrich 🆔
Universität zu Lübeck
Lübeck, Germany

ISSN 0302-9743 ISSN 1611-3349 (electronic)
Lecture Notes in Computer Science
ISBN 978-3-030-97280-6 ISBN 978-3-030-97281-3 (eBook)
https://doi.org/10.1007/978-3-030-97281-3

LNCS Sublibrary: SL6 – Image Processing, Computer Vision, Pattern Recognition, and Graphics

Preface

This volume comprises in total 27 scientific papers (18 long papers and nine short papers) that have all undergone peer review from the following three biomedical image analysis challenges at MICCAI 2021: the Mitosis Domain Generalization Challenge (MIDOG 2021), the Medical Out-of-Distribution Analysis Challenge (MOOD 2021), and Learn2Reg (L2R 2021). Our challenges share the need for developing and fairly evaluating algorithms that increase accuracy, reproducibility, and efficiency of automated image analysis in clinically relevant applications.

The MIDOG 2021 challenge aimed at finding domain-generic solutions for mitosis detection in histopathology images, a task commonly performed by pathologists in breast cancer diagnosis and grading. For this, 300 cases of breast cancer were digitized using six different microscopy scanners, two of which were completely unknown to the participants. Participants submitted dockered algorithms using the grand-challenge.org platform. The proceedings of MIDOG include five long papers followed by nine short papers.

The Learn2Reg competition's aim was to provide three complementary, clinically relevant tasks for medical image registration. Abdominal CT-MR fusion, respiratory motion estimation in CT, and whole-brain inter-subject alignment in MRI were addressed by a great variety of methods with considerable advances over previous state-of-the-art performance. The challenge was organized using the grand-challenge.org website as a point of contact for data sharing, submission of displacement fields and dockers, and evaluation.

With the MOOD 2021 challenge the goal was to provide a first standardized benchmark and challenge for out-of-distribution detection and localization on radiological imaging data. The challenge encompassed two publicly available training datasets and evaluation on the respective hidden test sets via docker submission and automated evaluation using the synapse.org platform.

The chairs of the organizing committees would like to express their sincere gratitude to the members of the organization committees and to the MICCAI challenges team.

January 2022

Marc Aubreville
MIDOG 2021 General Chair

David Zimmerer
MOOD 2021 General Chair

Mattias Heinrich
Learn2Reg 2021 General Chair

Contents

MOOD

L2R

MIDOG

MIDOG 2021 Preface

Algorithmic or algorithm-aided microscopy image processing has seen vast advancements in the last decade, thanks to powerful methods based on machine learning, and deep learning in particular. One task where automatic methods are especially helpful within the diagnostic process is the detection of cells undergoing cell division (mitotic figures) in digitized microscopy whole slide images. The density of mitotic figures within tumor tissue is known to be strongly correlated with tumor proliferation and is thus one of the strongest individual predictors for tumor malignancy in many tumor types, including breast cancer. While the determination of the count of mitotic figures within a predefined area is thus of great importance for prognostication, it is also notorious for having high inter-rater discordance and for being time consuming. Both call for the utilization of algorithms to increase diagnostic accuracy, reproducibility, and efficiency. Deep learning-based methods have shown to be remarkably accurate in the detection of mitotic figures recently, yet they often suffer from robustness issues when images from another laboratory are used for testing than for training. This drop in performance is caused by a domain-dependent covariate shift. Multiple factors contribute to domain shift, among which staining variability and tissue preparation differences have been long suspected to be influential factors. Another major factor, which has often been underestimated, is the optical and electrical properties of the image acquisition device (the whole slide scanner). Clinically applicable algorithmic solutions need to work with acceptable performance under a wide range of side conditions, including the scanner. Thus, they are required to generalize to unseen scanners.

This was the motivation for the first international Mitosis Domain Generalization Challenge (MIDOG 2021)[1], which was organized as a satellite event of the 24th International Conference on Medical Image Computing and Computer Assisted Intervention (MICCAI 2021). The goal of the challenge was to investigate and improve generalization across scanner types for mitosis detection, and thereby build the foundation for a general understanding of domain generalization for this task. The challenge was organized by a group of scientists from Austria, the Netherlands, and Germany. It opened for registration and download of the challenge dataset on April 1, 2021. As the target of the challenge was generalization and access to images of the test domains - even if unlabeled - could easily lead to compromised results, the participants were never able to download images of the test set. For the evaluation of the algorithms, they thus had to hand in containers encapsulating their computational approaches (using the docker virtualization framework) which were subsequently run by the organizers using an automated tool on the grand-challenge.org platform.

To check for algorithmic validity, the organizers made a preliminary test set available two weeks prior to the final submission deadline, and the participants were allowed to submit one container per day for a preliminary evaluation, albeit they received no access to the images themselves or the individual per-image results. As a baseline and for comparison, the organizers made available a reference approach by

---

[1] Official challenge description is available at: 10.5281/zenodo.4573978.

Wilm et al. together with an example docker container and a pre-print description of the approach. In total, 46 individuals submitted a container at this stage of the challenge, out of which 17 also submitted to the final test set. The deadline for the final submission was September 2, 2021.

Alongside the final submission, the participants had to provide a link to a publicly available pre-print short paper describing their approach in detail. All of these short papers were subjected to single-blind peer review, checking for novelty and quality of the work. Twelve approaches passed peer review and also surpassed a minimum score in the evaluation and were invited to participate in the workshop. One approach which passed peer review but did not exceed the minimum score was included as short paper in this proceedings. A long version of the description of the reference approach was subjected to external peer review by three experts and after acceptance added to the proceedings. All workshop participants were invited to also contribute a long paper to the proceedings, to be subjected to another round of peer review. In the end, five long papers and nine short papers were accepted to be included in the proceedings.

We thank all participants and the organization committee of MIDOG 2021 for their valuable contribution to the workshop, the discussions, and the proceedings.

Marc Aubreville
Katharina Breininger
Christof A. Bertram

MIDOG 2021 Organization

General Chair

Marc Aubreville   Technische Hochschule Ingolstadt, Germany

Program Committee

Katharina Breininger   Friedrich-Alexander-Universität
                                   Erlangen-Nürnberg, Germany
Christof A. Bertram   University of Veterinary Medicine, Austria
Nikolas Stathonikos   University Medical Center Utrecht,
                                   The Netherlands
Mitko Veta   Technical University Eindhoven,
                                   The Netherlands

Data and Technical Contributors

Robert Klopfleisch   Freie Universität Berlin, Germany
Natalie ter Hoeve   University Medical Center Utrecht,
                                   The Netherlands
Francesco Ciompi   Radboud University Medical Center,
                                   The Netherlands
Andreas Maier   Friedrich-Alexander-Universität
                                   Erlangen-Nürnberg, Germany

Domain Adversarial RetinaNet as a Reference Algorithm for the MItosis DOmain Generalization Challenge

Frauke Wilm[1]([✉]) [iD], Christian Marzahl[1] [iD], Katharina Breininger[2] [iD], and Marc Aubreville[3] [iD]

[1] Pattern Recognition Lab, Computer Sciences, Friedrich-Alexander-Universität, Erlangen-Nürnberg, Germany
frauke.wilm@fau.de
[2] Department of Artifical Intelligence in Biomedical Engineering, Friedrich-Alexander-Universität, Erlangen-Nürnberg, Germany
[3] Technische Hochschule Ingolstadt, Ingolstadt, Germany

Abstract. Assessing the mitotic count has a known high degree of intra- and inter-rater variability. Computer-aided systems have proven to decrease this variability and reduce labeling time. These systems, however, are generally highly dependent on their training domain and show poor applicability to unseen domains. In histopathology, these domain shifts can result from various sources, including different slide scanning systems used to digitize histologic samples. The MItosis DOmain Generalization challenge focused on this specific domain shift for the task of mitotic figure detection. This work presents a mitotic figure detection algorithm developed as a baseline for the challenge, based on domain adversarial training. On the challenge's test set, the algorithm scored an F_1 score of 0.7183. The corresponding network weights and code for implementing the network are made publicly available.

Keywords: MIDOG · Domain Shift · Mitotic Count · Histopathology · Object Detection

1   Introduction

A well-established method of assessing tumor proliferation is the mitotic count (MC) [12] - a quantification of mitotic figures in a selected field of interest. Identifying mitotic figures, however, is prone to a high level of intra- and inter-observer variability [3]. Recent work has shown that deep learning-based algorithms can guide pathologists during MC assessment and lead to faster and more accurate results [3]. However, these algorithmic solutions are highly domain-dependent and performance significantly decreases when applying these algorithms to data from unseen domains [7]. In histopathology, domain shifts are often attributed to varying sample preparation or staining protocols used at different laboratories. These domain shifts and their impact on the resulting performance of an

© Springer Nature Switzerland AG 2022
M. Aubreville et al. (Eds.): MIDOG 2021/MOOD 2021/L2R 2021, LNCS 13166, pp. 5–13, 2022.
https://doi.org/10.1007/978-3-030-97281-3_1

algorithm have been tackled with a wide range of strategies, e.g. stain normalization [9], stain augmentation [14], and domain adversarial training [7]. Domain shifts, however, cannot only be attributed to staining variations but can also include variations induced by different slide scanners [2]. The MItosis DOmain Generalization (MIDOG) challenge [1], hosted as a satellite event of the 24th International Conference on Medical Image Computing and Computer Assisted Intervention (MICCAI) 2021, addresses this topic in the form of assessing the MC on a multi-scanner dataset. This work presents the reference algorithm developed out-of-competition as a baseline for the MIDOG challenge. The RetinaNet-based architecture was trained in a domain adversarial fashion and scored an F_1 score of 0.7183 on the final test set.

2 Materials and Methods

The reference algorithm was developed on the official training subset of the MIDOG dataset [4]. We did not use any additional datasets and had no access to the (preliminary) test set during method development. The algorithm is based on a publicly available implementation of RetinaNet [10] which was extended by a domain classification path to enable domain adversarial training.

2.1 Dataset

The MIDOG training subset consists of Whole Slide Images (WSIs) from 200 human breast cancer tissue samples stained with routine Hematoxylin & Eosin (H&E) dye. The samples were digitized with four slide scanning systems: the Hamamatsu XR, the Hamamatsu S360, the Aperio CS2, and the Leica GT450, resulting in 50 WSIs per scanner. For the slides of three scanners, a selected field of interest sized approximately $2\,mm^2$ (equivalent to ten high power fields) was annotated for mitotic figures and hard negative look-alikes. These annotations were collected in a multi-expert blinded set-up. Aiming to support unsupervised domain adaptation approaches, no annotations were available for the Leica GT450 so that participants could only use the images for learning a visual representation of the scanner. Figure 1 illustrates exemplary patches of the scanners included in the training set.

The preliminary test set consists of five WSIs each for four slide scanning systems: the Hamamatsu XR and the Leica GT450, which already contributed to the training set, and the 3DHISTECH PANNORAMIC 1000 and the Hamamatsu RS, which were not seen during training. The scanner models of the preliminary test set, however, were undisclosed for the duration of the challenge. Participants only knew that the preliminary test set consisted of two seen and two unseen domains. This preliminary test set was used for evaluating the algorithms before submission and publishing preliminary results on a leaderboard on Grand Challenge[1]. The evaluation on Grand Challenge ensured that the participants had no access to test images during method development. This restriction

---

[1] https://midog2021.grand-challenge.org/.

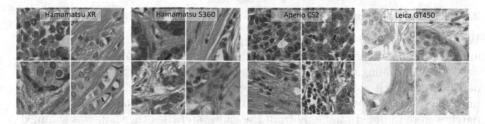

Fig. 1. Exemplary patches from the MItosis DOmain Generalization (MIDOG) challenge. Figure reproduced with permission from Aubreville *et al.* [2].

was also followed for developing the baseline algorithm. The final test set consists of 20 additional WSIs from the same scanners used for the preliminary test set. After the submission deadline, all algorithms were deployed once on this final test set for method comparison.

2.2   Domain Adversarial RetinaNet

For the domain adversarial training, we customized a publicly available RetinaNet implementation [10] by adding a Gradient Reversal Layer (GRL) and a domain classifier. For the encoder, we used a ResNet18 backbone pre-trained on ImageNet. For the domain discriminator, we were inspired by the work of Pasqualino *et al.* [13] and likewise chose three repetitions of a sequence of a convolutional layer, batch normalization, ReLU activation, and Dropout, followed by an adaptive average pooling and a fully connected layer. Implementation details can be obtained from our GitHub repository. We experimented with varying the number and positions of the domain classifier but ultimately decided for positioning a single discriminator at the bottleneck of the encoding branch. Figure 2 schematically visualizes the modified architecture.

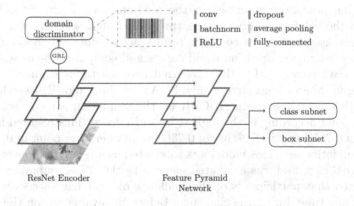

Fig. 2. Domain adversarial RetinaNet architecture.

2.3   Network Training

We split our training data into 40 training and ten validation WSIs per scanner and ensured a similar distribution of high and low MC samples in each subset. For network training, we used a patch size of 512×512 pixels and a batch size of 12. Each batch contained three images of each scanner. To overcome class imbalance, we employed a custom patch sampling, where half of the training patches were sampled randomly from the slides and the other half was sampled in a 512-pixel radius around a randomly chosen mitotic figure. Furthermore, we performed online data augmentation with random flipping, affine transformations, and random lightning and contrast change. The loss was computed as the sum of the domain classification loss for all scanners and the bounding box regression and instance classification loss for all annotated scanners:

$$\mathcal{L} = \sum_{s \in S} \frac{1}{M_s} \sum_{m=1}^{M_s} \mathcal{L}_{\text{dom},m} + \beta(s) \cdot (\mathcal{L}_{\text{bb},m} + \mathcal{L}_{\text{inst},m}) \quad \beta(s) = \begin{cases} 0, & \text{if } s = \text{GT450.} \\ 1, & \text{otherwise.} \end{cases}$$

S : set of scanners M : samples in batch

The bounding box loss $\mathcal{L}_{\text{bb}}$ was computed as smooth L1 loss and the focal loss [8] function was used for both, the instance ($\mathcal{L}_{\text{inst}}$) and the domain ($\mathcal{L}_{\text{dom}}$) classification loss. During backpropagation, the gradient was negated by the GRL and multiplied with α, a weighting factor which was gradually increased from 0 to 1 following the exponential update scheme of Ganin *et al.* [6]. We trained the network with a cyclical maximal learning rate of 1×10^{-4} for 200 epochs until convergence. Model selection was guided by the highest performance on the validation set as well as the highest domain confusion, i.e. highest domain classification loss, to ensure domain independence of the computed features.

2.4   Evaluation

The training procedure described in the previous section was repeated three times and the validation slides of the three annotated scanners were used for performance assessment. To compare results across different model operating points, we constructed precision-recall curves and compared the area under the precision-recall curves (AUCPRs) averaged over all three scanners for which mitotic figure annotations were available. As our final model, we selected the model with the highest mean AUCPR on the validation set and selected the operating point according to the highest mean F_1 score. This resulted in a mean AUCPR of 0.7551 and an F_1 score of 0.7369 at an operating point of 0.64 on our internal validation set. This model was submitted as a reference approach to the MIDOG challenge and was evaluated using a Docker-based submission system that ensured that participants of the challenge did not have access to the test images at any time during the challenge. Before the evaluation on the final test set, we ensured the sanity of the baseline algorithm by applying the model to the preliminary test set, which resulted in an F_1 score of 0.7401. This evaluation was run once, i.e., no hyperparameters were tuned on the preliminary test set.

For quantitative evaluation, we computed the F_1 score for mitosis detection on the challenge test set and compared the performance of the "reference approach", trained with domain adversarial training, to a weak baseline" trained without normalization or augmentation and a "strong baseline" trained with normalized images and the same online data augmentation methods as described in Sect. 2.3 but without methods for domain adaptation.

3 Results and Discussion

Across all test images, our weak baseline scored an F_1 score of 0.6279, our strong baseline an F_1 score of 0.6982, and our reference approach an F_1 score of 0.7183. Detailed results for precision, recall, and F_1 scores of the three models by scanner are summarized in Table 1. They show that the improved F_1 score over the strong baseline could mainly be attributed to a higher recall, i.e. less mitotic figures were overlooked, while precision values were very similar for most scanners.

Table 1. Performance metrics per model and scanner. The Hamamatsu XR also contributed to the training set with labeled images and the Leica GT450 with unlabeled images. The other scanners were unseen during training.

	Precision			Recall			F_1 score		
	Weak Baseline	Strong Baseline	Reference Approach	Weak Baseline	Strong Baseline	Reference Approach	Weak Baseline	Strong Baseline	Reference Approach
Seen Domains									
XR	**0.8043**	0.7778	0.7678	0.7291	0.7586	**0.7980**	0.7649	**0.7681**	0.7183
GT450	**0.9016**	0.7360	0.7318	0.2792	**0.6650**	**0.6650**	0.4264	**0.6987**	0.6968
Unseen Domains									
PANNORAMIC 1000	0.6698	0.5692	**0.6723**	0.7172	0.7475	**0.8081**	0.6927	0.6463	**0.7339**
RS	**0.6559**	0.6417	0.6364	0.4919	**0.6210**	**0.6210**	0.5622	**0.6311**	0.6286
All Scanners	**0.7545**	0.6965	0.7143	0.5377	0.6998	**0.7223**	0.6279	0.6982	**0.7183**

In Fig. 3a, we used bootstrapping to visualize the distribution of F_1 scores per scanner. The results show that the weak baseline performed particularly badly for the Leica GT450 scanner with an average F_1 score of 0.4264 and a high variance in performance across all test slides, which becomes apparent by the wide distribution in the bootstrapping visualization. Looking at the detailed results in Table 1, this was mainly attributed to a low recall, i.e. a lot of mitotic figures were overlooked. Considering the example patches of the Leica scanner shown in Fig. 1, this result is not surprising, as the Leica scanner produces images with a much higher illumination and less contrast compared to the other scanners. Without normalization, these images can challenge the network, especially since the Leica scanner was not seen during training of the baseline models due to missing annotations and was only used for training the domain generalization component of the domain adversarial network. When comparing the strong baseline with our reference approach, the models show very similar performance for most of the scanners except for the unseen PANORAMIC 1000, where the domain adversarial training significantly increased the F_1 score to 0.7339 compared to an F_1 score of 0.6463 for the strong baseline. Furthermore, the narrower distributions of the bootstrapping in Fig. 3d indicate a lower variance in performance compared to the wider distributions of the baseline models in Fig. 3b and c.

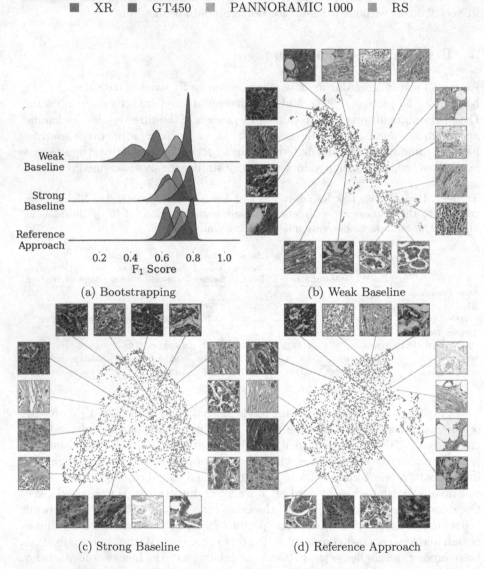

Fig. 3. Bootstrapping and Uniform Manifold Approximation and Projection (UMAP) plots of the evaluated models. The weak baseline was trained without any measures for normalization or augmentation and the strong baseline was trained with normalized images and online augmentations.

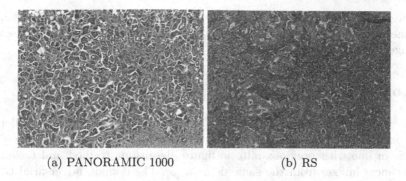

(a) PANORAMIC 1000   (b) RS

Fig. 4. Exemplary images where F_1 scores for the strong baseline and the domain adversarial varied significantly.

Additionally, we evaluated the models' capability for domain generalization by using Uniform Manifold Approximation and Projection (UMAP) [11] plots. UMAP is a dimensionality reduction technique that can be used to visualize the high dimensional feature representations within neural networks in a two-dimensional space. For our plots, we have randomly sampled 30 patches on each WSI of the MIDOG test set and selected the output of the last layer of our RetinaNet encoders for visualization. The UMAP plot of the reference approach is visualized in Fig. 3d. The data clustering independent of scanner domains shows that the domain adversarial training encouraged the extraction of domain-independent features. As a comparison Fig. 3b visualizes the UMAP plot for the weak baseline. Here, the samples show a distinctive clustering according to scanner vendors. The cluster centers of the two Hamamatsu scanners are closer together, which is not surprising as they come from the same vendor and the same series (NanoZoomer). Figure 3c shows the UMAP plot of the strong base-line. Whereas the normalization and augmentation techniques pushed the distributions closer together, the GT450 still forms a distinguishable cluster at the lower right of the feature representation. Recalling the scanner-wise model performance summarized in Table 1, however, this did not impair the mitosis detection. Nevertheless, when comparing the bootstrapping visualizations in Fig. 3c and d, the remaining three scanners are less distinguishable in the feature representation of the domain adversarial model which seemed to have helped the mitotic figure detection for especially the unseen scanners. Interestingly, Fig. 3d shows a separated cluster on the right hand of the main cluster with patches from all scanners. A closer look at the example patches shows that these were predominantly patches with large white areas due to teared tissue or empty fat vacuoles.

Figure 4 shows two examples where the domain adversarial model significantly outperformed the strong baseline with F_1 scores of 0.8 and 0.6 for the PANORAMIC 1000 image in Fig. 4a and F_1 scores of 0.6364 and 0.4286 for the Hamamatsu RS image in Fig. 4b. The large differences in performance could mainly be attributed to a higher number of false-positive predictions for the baseline model. Both examples show very intense staining which might not have been met with the augmentation methods used during training and thereby challenged the strong baseline model.

4   Conclusion

In this work, we presented our baseline algorithm for the MIDOG challenge, based on domain adversarial training. With an F_1 score of 0.7183, the algorithm is in line with previous mitotic figure algorithms trained and tested on breast cancer images from the same domain [5]. The domain adversarial training improved especially the generalization across unseen scanner domains while maintaining a similar performance on seen domains. The feature representation as UMAP plots visualizes the successful extraction of domain invariant features of the proposed network. In total, 17 algorithms were submitted to the MIDOG challenge for evaluation on the final test set. From these, four approaches outperformed this strong but out-of-competition reference approach. The code used for implementing and training the proposed network is publicly available in our GitHub[2] repository.

References

1. Aubreville, M., et al.: Mitosis domain generalization challenge. Zenodo (2021). https://doi.org/10.5281/zenodo.4573978
2. Aubreville, M., et al.: Quantifying the scanner-induced domain gap in mitosis detection. In: Medical Imaging with Deep Learning (2021)
3. Aubreville, M., et al.: Deep learning algorithms out-perform veterinary pathologists in detecting the mitotically most active tumor region. Sci. Rep. **10**(16447), 1–11 (2020)
4. Aubreville, M., et al.: MItosis DOmain Generalization Challenge (MICCAI-MIDOG 2021) Training Data. Zenodo, April 2021. https://doi.org/10.5281/zenodo.4643381
5. Bertram, C.A., et al.: Are pathologist-defined labels reproducible? Comparison of the TUPAC16 mitotic figure dataset with an alternative set of labels. In: Cardoso, J., et al. (eds.) IMIMIC/MIL3ID/LABELS -2020. LNCS, vol. 12446, pp. 204–213. Springer, Cham (2020). https://doi.org/10.1007/978-3-030-61166-8_22
6. Ganin, Y., Lempitsky, V.: Unsupervised domain adaptation by backpropagation. In: International Conference on Machine Learning, pp. 1180–1189. PMLR (2015)
7. Lafarge, M.W., Pluim, J.P.W., Eppenhof, K.A.J., Moeskops, P., Veta, M.: Domain-adversarial neural networks to address the appearance variability of histopathology images. In: Cardoso, M.J., et al. (eds.) DLMIA/ML-CDS -2017. LNCS, vol. 10553, pp. 83–91. Springer, Cham (2017). https://doi.org/10.1007/978-3-319-67558-9_10
8. Lin, T.Y., Goyal, P., Girshick, R., He, K., Dollár, P.: Focal loss for dense object detection. In: Proceedings of the IEEE International Conference on Computer Vision, pp. 2980–2988 (2017)

---

[2] https://github.com/DeepPathology/MIDOG.

9. Macenko, M., et al.: A method for normalizing histology slides for quantitative analysis. In: 2009 IEEE International Symposium on Biomedical Imaging: From Nano to Macro, pp. 1107–1110. IEEE (2009)
10. Marzahl, C., et al.: Deep learning-based quantification of pulmonary hemosiderophages in cytology slides. Sci. Rep. **10**(1), 1–10 (2020)
11. McInnes, L., Healy, J., Melville, J.: Umap: uniform manifold approximation and projection for dimension reduction (2020)
12. Meuten, D., Moore, F., George, J.: Mitotic count and the field of view area: time to standardize (2016)
13. Pasqualino, G., Furnari, A., Signorello, G., Farinella, G.M.: An unsupervised domain adaptation scheme for single-stage artwork recognition in cultural sites. Image Vis. Comput. **107**, 104098 (2021)
14. Tellez, D., Balkenhol, M., Karssemeijer, N., Litjens, G., van der Laak, J., Ciompi, F.: H and E stain augmentation improves generalization of convolutional networks for histopathological mitosis detection. In: Medical Imaging 2018: Digital Pathology, vol. 10581, p. 105810Z. International Society for Optics and Photonics (2018)

Assessing Domain Adaptation Techniques for Mitosis Detection in Multi-scanner Breast Cancer Histopathology Images

Jack Breen[1]($\boxtimes$) (iD), Kieran Zucker[2] (iD), Nicolas M. Orsi[2] (iD),
and Nishant Ravikumar[1] (iD)

[1] CISTIB Center for Computational Imaging and Simulation Technologies
in Biomedicine, School of Computing, University of Leeds, Leeds, UK
{scjjb,N.Ravikumar}@leeds.ac.uk
[2] Leeds Institute of Medical Research at St James's, School of Medicine,
University of Leeds, Leeds, UK

Abstract. Breast cancer is the most commonly diagnosed cancer worldwide, with over two million new cases each year. During diagnostic tumour grading, pathologists manually count the number of dividing cells (mitotic figures) in biopsy or tumour resection specimens. Since the process is subjective and time-consuming, data-driven artificial intelligence (AI) methods have been developed to automatically detect mitotic figures. However, these methods often generalise poorly, with performance reduced by variations in tissue types, staining protocols, or the scanners used to digitise whole-slide images. Domain adaptation approaches have been adopted in various applications to mitigate this issue of domain shift. We evaluate two unsupervised domain adaptation methods, CycleGAN and Neural Style Transfer, using the MIDOG 2021 Challenge dataset. This challenge focuses on detecting mitotic figures in whole-slide images digitised using different scanners. Two baseline mitosis detection models based on U-Net and RetinaNet were investigated in combination with the aforementioned domain adaptation methods. Both baseline models achieved human expert level performance, but had reduced performance when evaluated on images which had been digitised using a different scanner. The domain adaptation techniques were each found to be beneficial for detection with data from some scanners but not for others, with the only average increase across all scanners being achieved by CycleGAN on the RetinaNet detector. These techniques require further refinement to ensure consistency in mitosis detection.

Keywords: Convolutional Neural Network (CNN) · Generative Adversarial Network (GAN) · Neural Style Transfer · CycleGAN

1 Introduction

Breast cancer is the most commonly diagnosed cancer worldwide, accounting for one quarter of all malignancies in women [20]. Diagnosis involves identifying

© Springer Nature Switzerland AG 2022
M. Aubreville et al. (Eds.): MIDOG 2021/MOOD 2021/L2R 2021, LNCS 13166, pp. 14–22, 2022.
https://doi.org/10.1007/978-3-030-97281-3_2

cancer sub-type, grade, and molecular profile (oestrogen/progesterone receptor and HER2 amplification status). Counting dividing cells (mitotic figures) is a key task for pathologists within the Nottingham grading score, which combines mitotic count, tubule formation, and nuclear pleomorphism as a measure of the aggressiveness of the underlying malignancy. This grade carries prognostic information as part of the Nottingham Prognostic Index [21].

Recent studies in human and veterinary pathology have highlighted that pathologists' detection of mitoses is both variable [3] and time-consuming [9]. This increases pressure on histopathology diagnostic services, where in the UK alone, only 3% of departments are adequately staffed to meet diagnostic demand. As a result, 45% of departments routinely outsource work and 50% use locums [18], at significant cost. Recent advances in deep learning (DL)-driven automated techniques for mitosis detection have shown promise both for reducing inter-observer variability between pathologists, and for relieving some of the workload associated with the mitotic count.

Domain adaptation approaches in general aim to learn a mapping that reduces the gap between source and target data distributions. In the context of computer vision problems, they are employed to improve generalisation of image-based DL models to data from different domains during inference. While convolutional neural networks (CNNs) have proved a powerful tool for solving a multitude of vision problems, it is well established that they tend to over-fit to data in the training domain, and hence generalise poorly to target domains during inference. Domains in digital histopathology may include variations introduced from tissue staining processes, scanner properties, or the histological preparations being scanned. The domain adaptation methods we investigate focus on the visual appearance of an image, taking a content image and one or more style images, and creating a stylised representation of the content image.

1.1 MIDOG Challenge 2021 and Relevant Literature

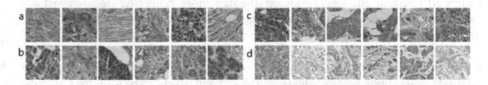

Fig. 1. 512 × 512 segments of MIDOG 2021 training data from (a) Hamamatsu XR (HXR), (b) Hamamatsu S360 (HS360), (c) Aperio CS (ACS), (d) Leica GT.

The MIDOG Challenge was a competition on mitosis detection [1], held at the International Conference on Medical Image Computing and Computer Assisted Intervention, 2021. The challenge provided 200 training images, each a $2\,mm^2$ region of interest selected manually from haematoxylin and eosin (H&E) stained breast cancer whole-slide images (WSIs). All samples were prepared using an identical staining process at the pathology laboratory of UMC Utrecht. These images were digitised using four different scanners: 50 each from Hamamatsu XR nanozoomer 2.0 (HXR), Hamamatsu S360 0.5 NA (HS360), Aperio ScanScope CS2 (ACS), and Leica GT450. All images digitised using the first three scanners were annotated via majority voting by three pathologists, to reduce inter-observer variability and single-observer bias. A deep learning model was used to suggest any potentially missed mitotic figures, which were also annotated by the pathologists. Scans from the Leica GT450 lacked annotations and were not used in this study. Algorithms submitted to the challenge were evaluated on a set of 80 WSIs from a combination of training set scanners and previously unseen scanners.

Efforts were made to minimise selection bias during data collection for the challenge, by digitising all available breast cancer samples in the order that they had been sent to the pathology lab for examination. This led to a highly varied training set, with the number of annotations for a WSI ranging from 2 to 184. These annotations were sparsely distributed throughout the images, with most 512×512 crops containing no annotations at all. The difference between scanners is shown in Fig. 1. The scanners have different colour profiles, with ACS producing deeper red scans, and HXR producing scans with a blue-purple hue.

Prior to MIDOG 2021, few studies had investigated the impact of domain adaptation techniques on the detection of mitotic figures in images acquired using multiple scanners. A recent study used a domain adversarial neural network (DANN), a multi-task approach which combines domain adaptation and classification to learn image domain-independent representations [16]. However, the method performed poorly when combined with stain normalisation. This suggests that the model learned to classify within specific domains rather than truly learning to classify in a domain-independent manner.

Another study proposed a stain-transfer network using a GAN architecture with a U-Net encoder, adding an edge-weighted regularisation to retain basic structures from the input images [2]. This was found to improve performance for patch-level classification, but was not tested on pixel-level predictions for localising and counting mitoses. In histopathology more generally, researchers have evaluated the benefit of domain adaptation for segmentation [8], detection [10], and classification tasks [19].

The most common techniques are stain normalisations based upon matching colour distribution, and GANs for holistic domain adaptation. The domain adaptation methods we investigate in the current study have a limited amount of previous research in histopathology. Two studies investigated Neural Style Transfer (NST) for transferring stains across cell-level images, but without evaluating the subsequent effects on a computer vision model [4,7]. To the best of our knowledge, these are the only studies to use this method for histopathology without significant adjustments, such as changing to an adversarial loss function [15]. The original NST method has not been evaluated as a domain adaptation tool for cell-level classification, segmentation, or detection. CycleGAN is increasingly popular in histopathology [19], and while previous research in mitosis detection has evaluated it for switching between stains [14], it has not previously been evaluated for domain adaptation within the same staining process.

2 Methods

2.1 Mitosis Detection Models

U-Net is an architecture for semantic segmentation which combines multiple layers of downsampling to generate a multi-scale feature mapping [17]. This method requires segmentation masks for training, which we generated by taking each pixel to be a 1 if it was within a mitotic figure bounding box and a 0 otherwise. U-Net outputs a probability map, which we converted to bounding box predictions through a multi-step process, shown in Fig. 2. First a binary map was generated by applying a threshold to the probability map. Objects were subsequently extracted from this by selecting external contours. Any detection with a height or width less than 10 pixels was assumed to be an artifact and was removed, as this was empirically found to improve robustness. The remaining detections had a bounding box placed around their center at the same size as the original annotations. We used a combination of binary cross entropy loss, dice loss and focal loss, which we weighted heavily towards the focal loss as this performs well on unbalanced datasets [11].

RetinaNet is a one-stage detection algorithm which feeds inputs through an encoder and a feature pyramid network to generate multi-scale features. These features are used for simultaneous bounding box regression and classification [11]. This has previously been shown to perform at the level of an expert pathologist for quantifying pulmonary haemosiderophages [13].

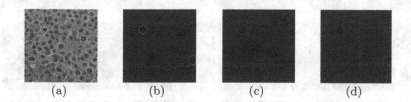

(a)   (b)   (c)   (d)

Fig. 2. U-Net post-processing procedure. (a) ground-truth image with one mitotic figure (b) predictions map output from U-Net (c) binary map (d) bounding box predictions. This example has one True Positive prediction (upper left) and one False Positive prediction (lower right).

2.2 Domain Adaptation Methods

Neural Style Transfer (NST) is a one-to-one domain adaptation method, casting the style of one image on to the content of another [5]. NST uses intermediate layers of a pre-trained VGG19 CNN to extract features representing the style and content of each input image. The loss function combines a style loss, which quantifies the difference between the output image and the input style image, and a content loss, which quantifies the difference between the output and the input content images.

CycleGAN is a generative adversarial network used for visual domain adaptation [22]. CycleGAN uses two GANs, one to produce a stylised image and another to recreate the original input image from the stylised image. This attempts to overcome mode collapse, where a generator creates the same output regardless of input. The performance at reproducing the original input image is measured by a cycle consistency loss, and the performance at transferring the style is measured by an adversarial loss. As GAN losses tend not to give a clear indication of convergence, we use the Fréchet Inception Distance to decide when to stop training [6].

Macenko normalisation is a common normalisation approach in H&E-stained images which accounts for each stain separately [12]. This was used for comparison to the domain adaptation approaches. The normalisation was applied to both the training data and the evaluation data, where the domain adaptation approaches were only applied to evaluation data.

Examples of all three methods are shown in Fig. 3. Neural Style Transfer changes the colour profile of scans much less than CycleGAN, as it was found that running NST for more iterations led to very poor mitosis detection performance, as the resulting images were too artificial.

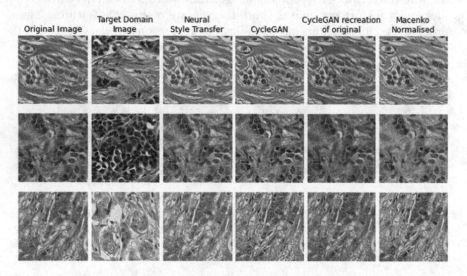

Fig. 3. Domain Adaptation and normalisation approaches applied to three crops from the Hamamatsu S360 scanner, with Aperio CS as the target domain.

2.3   Implementation

Models were evaluated using the cell-wise F1 score, which is the harmonic mean of the precision and recall. This metric punishes false detections and missed mitotic figures equally. True positives are defined as predictions made within 30 pixels of the center of an annotated mitotic figure. Both domain adaptation methods were used as normalisation approaches during inference, not during training. This approach keeps the detection models completely agnostic to the testing domain, ensuring that any difference to detection performance is a result of the domain adaptation approaches alone.

All experiments were undertaken with a single GPU on Google Colab. Both of the detectors were implemented with a ResNet encoder pre-trained on ImageNet, with ResNet101 found to be optimal for RetinaNet and ResNet152 for U-Net. We evaluated our methods using a three-fold approach where two annotated scanners were used for training and the other one was withheld for evaluation as an unseen domain. The average F1 score and 95% confidence intervals were calculated using 10,000 epoch bootstrapping, with evaluations on non-overlapping 512×512 crops from all available WSIs in the training domain (40 training WSIs and 10 test WSIs per scanner), and from all 50 WSIs in the external domain.

3   Results

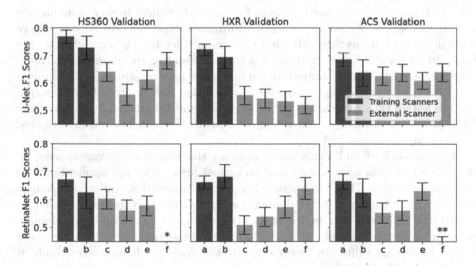

Fig. 4. F1 scores from each detection model, where two scanners were used for training and one for validation, on (a) the training set (b) a test set from the training domain (c) the external domain (d) the external domain with NST (e) the external domain with CycleGAN (f) the external domain with Macenko. 95% confidence intervals shown in black. *0.243 ± 0.043. **0.426 ± 0.041.

Mitosis detection results from all experiments conducted in this study are summarised as bar plots of F1 scores in Fig. 4 with 95% confidence intervals. The best average performance on unseen data was achieved by the U-Net with Macenko normalisation, with the baseline U-Net and U-Net+CycleGAN close behind. At the time of MIDOG 2021, our best results were from U-Net without domain adaptation, so we trained this model on data from all three training scanners. Our model achieved an F1 score of 0.693 on the preliminary test set, and 0.686 on the final test set. Precision and recall were very similar, at 0.686 and 0.685, respectively, indicating that the model was balanced.

4   Discussion

The performance of both baseline detection algorithms is found to be comparable to human performance, with an average F1 score of 0.69 for U-Net and 0.65 for RetinaNet, compared to the human score of 0.68 in a comparable study [3]. The average performance on the unseen domain is lower, at 0.61 and 0.56, respectively. Detection performance was generally lower when the Hamamatsu XR was used as an unseen scanner, which is likely caused by the scanner having a significantly different colour profile to the other two scanners. Both domain adaptation methods slightly improved performance on the RetinaNet, though this performance was still lower than on the other scanners.

Overall, the domain adaptation methods performed inconsistently, with CycleGAN improving the average F1 score for RetinaNet but not for U-Net, and NST improving the average score for one of the three unseen domains for each detector, but degrading average performance. Macenko normalisation gave a slight increase in average detection performance for the U-Net, but was very inconsistent for RetinaNet, with both domain adaptation techniques performing better on average. Due to computational limits and long run-times for the domain adaptation methods, hyperparameter optimization was limited. Our hyperparameters were thus influenced by similar works and by practicalities, and were likely to be sub-optimal for mitotic figure detection, which was not investigated in these similar works.

Future work should focus on evaluating the source of inconsistencies in domain adaptation methods applied to digital histopathology applications, to improve reliability. This may include combining images from different scanners to create a more general target domain for CycleGAN. Furthermore, the effects of training both detection models with artificial images generated by the domain adaptation methods should be evaluated as this may make the detectors more robust to domain shifts. To better evaluate the domain adaptation methods, it would be beneficial to also compare their respective run-times.

5   Conclusion

We implemented[1] two unsupervised domain adaptation techniques, CycleGAN and Neural Style Transfer, for overcoming scanner-driven domain shifts in histology images, and enabling robust mitosis detection. These were applied to

---

[1] https://github.com/scjjb/MIDOG_Domain_Adaptation.

transform test data acquired using scanners previously unseen by the detection models. Our baseline detection methods, U-Net and RetinaNet, performed comparably to a human expert on data from the training domain, with reduced performance on data from an unseen domain. Both domain adaptation techniques were found to improve detection performance for some of the unseen domains but not for all. Average detection performance across all unseen domains was only improved using CycleGAN in combination with the RetinaNet detector. This partial success justifies the need for further investigation to understand and overcome these inconsistencies. Improved techniques for modelling in the presence of domain-shifts, or for learning domain-invariant features, will be essential for accelerating the deployment and adoption of automated tools in routine diagnostic practice.

References

1. Aubreville, M., et al.: Mitosis domain generalization challenge (2021). https://doi.org/10.5281/zenodo.4573978
2. BenTaieb, A., Hamarneh, G.: Adversarial stain transfer for histopathology image analysis. IEEE Trans. Med. Imaging **37**, 792–802 (2018). https://doi.org/10.1109/TMI.2017.2781228
3. Bertram, C.A., et al.: Computer-assisted mitotic count using a deep learning-based algorithm improves inter-observer reproducibility and accuracy in canine cutaneous mast cell tumors (2021). https://doi.org/10.1101/2021.06.04.446287
4. Ganesh, A., Vasanth, N.R., George, K.: Staining of histopathology slides using image style transfer algorithm (2019). https://doi.org/10.1109/SSCI.2018.8628672
5. Gatys, L., Ecker, A., Bethge, M.: A neural algorithm of artistic style. J. Vis. **16** (2016). https://doi.org/10.1167/16.12.326
6. Heusel, M., Ramsauer, H., Unterthiner, T., Nessler, B., Hochreiter, S.: GANs trained by a two time-scale update rule converge to a local nash equilibrium, vol. 2017-December (2017)
7. Izadyyazdanabadi, M., et al.: Fluorescence image histology pattern transformation using image style transfer. Front. Oncol. **9** (2019). https://doi.org/10.3389/fonc.2019.00519
8. Khan, A.M., Rajpoot, N., Treanor, D., Magee, D.: A nonlinear mapping approach to stain normalization in digital histopathology images using image-specific color deconvolution. IEEE Trans. Biomed. Eng. **61**, 1729–1738 (2014). https://doi.org/10.1109/TBME.2014.2303294
9. Laflamme, P., et al.: Phospho-histone-H3 immunostaining for pulmonary carcinoids: impact on clinical appraisal, interobserver correlation, and diagnostic processing efficiency. Hum. Pathol. **106**, 74–81 (2020). https://doi.org/10.1016/j.humpath.2020.09.009
10. Liimatainen, K., Kananen, L., Latonen, L., Ruusuvuori, P.: Iterative unsupervised domain adaptation for generalized cell detection from brightfield z-stacks. BMC Bioinform. **20**, 1–10 (2019). https://doi.org/10.1186/s12859-019-2605-z
11. Lin, T.Y., Goyal, P., Girshick, R., He, K., Dollar, P.: Focal loss for dense object detection. IEEE Trans. Pattern Anal. Mach. Intell. **42** (2020). https://doi.org/10.1109/TPAMI.2018.2858826
12. Macenko, M., et al.: A method for normalizing histology slides for quantitative analysis (2009). https://doi.org/10.1109/ISBI.2009.5193250

13. Marzahl, C., et al.: Deep learning-based quantification of pulmonary hemosiderophages in cytology slides. Sci. Rep. **10**, 9795 (2020). https://doi.org/10.1038/s41598-020-65958-2
14. Mercan, C., et al.: Virtual staining for mitosis detection in breast histopathology. In: 2020 IEEE 17th International Symposium on Biomedical Imaging (ISBI), pp. 1770–1774 (2020). https://doi.org/10.1109/ISBI45749.2020.9098409
15. Nishar, H., Chavanke, N., Singhal, N.: Histopathological stain transfer using style transfer network with adversarial loss. In: Martel, A.L., et al. (eds.) MICCAI 2020. LNCS, vol. 12265, pp. 330–340. Springer, Cham (2020). https://doi.org/10.1007/978-3-030-59722-1_32
16. Otálora, S., Atzori, M., Andrearczyk, V., Khan, A., Müller, H.: Staining invariant features for improving generalization of deep convolutional neural networks in computational pathology. Front. Bioeng. Biotechnol. **7** (2019). https://doi.org/10.3389/fbioe.2019.00198
17. Ronneberger, O., Fischer, P., Brox, T.: U-Net: convolutional networks for biomedical image segmentation. In: Navab, N., Hornegger, J., Wells, W.M., Frangi, A.F. (eds.) MICCAI 2015. LNCS, vol. 9351, pp. 234–241. Springer, Cham (2015). https://doi.org/10.1007/978-3-319-24574-4_28
18. Royal College of Pathologists: Meeting pathology demand: Histopathology workforce census (2018)
19. Shin, S.J., et al.: Style transfer strategy for developing a generalizable deep learning application in digital pathology. Comput. Methods Programs Biomed. **198** (2021). https://doi.org/10.1016/j.cmpb.2020.105815
20. Sung, H., et al.: Global cancer statistics 2020: globocan estimates of incidence and mortality worldwide for 36 cancers in 185 countries. CA Cancer J. Clin. **71**, 209–249 (2021). https://doi.org/10.3322/caac.21660
21. World Health Organization: WHO Classification of Breast Tumours, vol. 2 (2019)
22. Zhu, J.Y., Park, T., Isola, P., Efros, A.A.: Unpaired image-to-image translation using cycle-consistent adversarial networks (2017). https://doi.org/10.1109/ICCV.2017.244

Domain-Robust Mitotic Figure Detection with Style Transfer

Youjin Chung[iD], Jihoon Cho[iD], and Jinah Park[✉][iD]

Korea Advanced Institute of Science and Technology, Daejeon, South Korea
{krista2811,zinic,jinahpark}@kaist.ac.kr

Abstract. Recent studies for mitotic figure identification have shown performance comparable to that of human experts; however, the challenge to develop strategies invariant to image variance in different microscope slide scanners still remains. In this paper, we propose a method for domain generalization in mitotic figure detection by considering the scanner as a *domain* and the characteristic specified for each domain as a *style*. The method aims to make the mitosis detection network robust to scanner types by using various style images. To expand the style variance, the style of the training image is transferred into arbitrary styles by the proposed style transfer module based on StarGAN. Furthermore, we propose patch selection criteria to resolve the imbalance problem. Our model with the proposed training scheme has obtained satisfactory detection performance on the MIDOG Challenge containing scanners that have not been seen.

Keywords: Domain Generalization · Mitosis Detection · Style Transfer

1 Introduction

Detecting mitotic figures (MFs) in histopathology images is important for tumor prognostication. Since counting the MFs by a pathologist is costly, computer-aided automatic MF detection is attracting more attention. Many research works [7,12,16] have achieved remarkable performance in detecting MF. However, microscopic images are obtained by various scanners and the images contain different visual features depending on the scanner, hence the works mentioned above show an inconsistent performance by different scanners. Focusing on the color difference between scanners, many research works [3,13] have proposed a method using stain normalization [10,14,19] with the principle of interpreting the problem as a domain generalization problem by defining the scanner as

This research is supported by the Ministry of Culture, Sports and Tourism and Korea Creative Content Agency (Project Number: R2020070002) and the Capacity Enhancement Program for Scientific and Cultural Exhibition Services through the National Research Foundation of Korea (NRF) funded by the Ministry of Science and ICT (NRF-2018X1A3A1069693).
Y. Chung and J. Cho contributed equally to this work as first authors.

M. Aubreville et al. (Eds.): MIDOG 2021/MOOD 2021/L2R 2021, LNCS 13166, pp. 23–31, 2022.
https://doi.org/10.1007/978-3-030-97281-3_3

the *domain*. However, stain normalization performs a simple pixel-by-pixel color mapping based on only the color distribution of images and some visual artifacts can occur [22].

We have considered some global aspects that capture overall context of the stained image to get better performance in MF detection. In this paper, we define a novel domain characteristic, namely *style*, which contains shape details as well as color distribution. Our proposed style transfer module (STM) converts domain style of an input image into a different style. We increase the number of training domains through STM so that the model learns the unique characteristic of MF regardless of the scanner. In addition, we devise the selection criteria of input patches to resolve the data imbalance issue. The effectiveness of our proposed method is explained in Sect. 4.2. The proposed method has achieved 0.7243 F1 score on the test set of the 2021 MICCAI MIDOG Challenge, ranking 4th in the leaderboard [1].

2   Related Work

Computer-aided mitosis detection has gained interests from researchers for the benefit of reducing the manual workload. The nature of the problem also fits in the current development of computing disciplines. A two-step method, feature extraction and classification method, was widely adapted by early approaches [7,12,16]. The feature extraction step is composed of preprocessing and generating candidate segmentation from the whole slide image (WSI). The selection of features sent to the next step is based on statistics [7,16] or prior knowledge of mitosis to reduce false negatives due to insufficient MF counts.

The emergence of deep learning and the development of computing power also has a large influence on the mitosis detection area. CNNs have shown great performance improvement in 2D image processing areas, so they have also been used in mitosis detection. Some approaches focus on input patches that are used in training [3,13] with stain normalization methods of [10,14,19]. Other approaches modify the model to be efficient and powerful [8,21]. However, the problem that remains for these approaches is that they have insufficient generalization to scanners. To solve the scanner dependency problem, Tellez et al. [17] used hematoxylin and eosin stain-based augmentation on input patches to devise a scanner-robust method.

Domain generalization is the problem of making a model robust to multidomain data containing an unseen test domain. Reducing the difference between domains is a widely used method for generalization [5,6,18], such as alignment of extracted features. Another approach is to augment the input data so that the model can be robust to unseen domains [15,23]. In this approach, augmentation is used as a solution for domain shift. Simple random augmentation and generative augmentation are existing approaches for augmentation. For generative domain shift, synthesis using a generative adversarial network (GAN) [4] is widely used. StarGAN [2] is known to perform well in multidomain environments, while some approaches [24] focus on two-domain situations. Our style transfer module uses StarGAN's generator, which is used to create plausible augmented patches by applying style transformation including both color and shape details. The module reflects the characteristics better than simple stain augmentation, which only changes the color distribution.

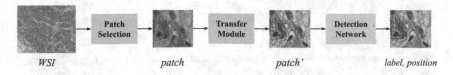

Fig. 1. Overview of our training scheme using the style transfer module.

3   Method

Our training scheme consists of a patch selection and a style transfer module. Patch selection is used to control the image patch used in the actual learning process. Then, the style transfer module changes the input patch to an arbitrary style. Finally, the detection network is trained by the transformed patch. Figure 1 shows the overall view of our proposed scheme.

3.1   Patch Selection (PS)

Performing the detection task using WSI as input has a high computational cost on both the STM and detection network. We use a patch-based approach that has been used in previous mitosis detection studies. Selection of the input patch is essential because the model performance is highly dependent on the input patches. We use several criteria to sort the input patches.

First, we consider the ratio of the foreground-background patches. The foreground (FG) patch denotes that it contains either MF or mitotic-like figure (MLF) annotation, and the background (BG) patch has neither type of annotation. When all slices of WSI are used in training, the model concentrates on the background because the portion of the annotated patch is small compared to BG patches. Therefore, we adjust the FG-BG ratio to $\alpha : 1$ such that BG patches have a small portion. Additionally, the number of FG patches is balanced between scanners to prevent the model bias of a certain scanner. We address the effect of our PS criteria in Sect. 4.2.

3.2   Style Transfer Module (STM)

The role of the style transfer module (STM) is to take an input patch and a random style code and return a generated patch that fits the style. When shifting between N domains, StarGAN uses an N-element one-hot vector in which 1 represents the target domain. Here, if this code is replaced with a normalized float vector, the generator should return an image with a new fused style.

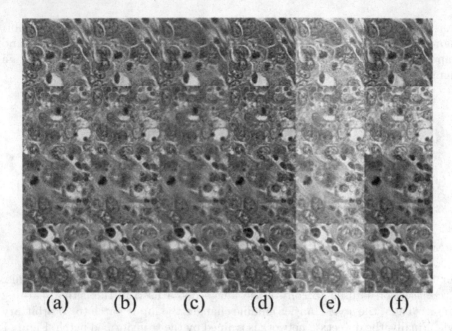

(a)   (b)   (c)   (d)   (e)   (f)

Fig. 2. Transfer module results. (a): original patch (b): scanner-1 style (c): scanner-2 style (d): scanner-3 style (e) scanner-4 style (f): random style

We follow the adversarial training scheme and loss function of Choi et al. [2] with the provided scanner images. Each scanner type represents a one-hot label of the style code vector. In the training step, the generator G is trained to transfer a scanner image into another scanner-styled image, while the discriminator D is trained to determine whether or not the image is synthesized. Using this approach, G will learn to transfer between scanner styles, including color distribution and shape details. Figure 2 (f) shows the generator making a random style when using the random normalized style code, as expected.

In training detection model, the generator of the trained StarGAN is used as our STM. STM performs augmentation before entering the detection network. Unlike the training stage, STM takes a normalized random float vector as a style code and makes the style-transferred image based on the corresponding code. Figure 3 shows the difference in training and inference steps. Furthermore, we modified the portion of the generated random styled image and the original patch image as p. This is because overgeneralization can degrade the overall performance.

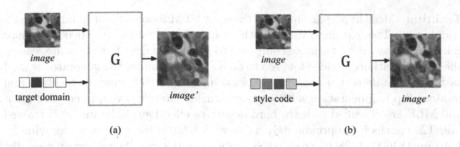

(a)   (b)

Fig. 3. Difference between StarGAN and style transfer modules. (a): StarGAN generator (b): Our proposed style transfer module (STM)

Moreover, we found some artifacts that hinder the quality of the generated image. There are two kinds of artifacts: checkerboards and stain artifacts. The checkerboard artifact is caused by trainable upsampling of transposed convolution. Following the solution introduced in the paper of [11], transposed convolution is changed to a stack of bilinear upsampling and convolution. Another problem is unexpected stain generation on the background, which might confuse the model. Therefore, by using the background segregation method from stain normalization by Macenko et al. [10], we separated and preserved the background part from generation. Since this approach preserves the style better than the previous version, we call this version advanced STM (a-STM).

4   Experiments

4.1   Implementation

The MIDOG Challenge dataset containing four scanners with approximately 7200×5400 pixel sized WSIs and 50 images for each scanner is used. Scanner annotations contain information on MF and MLF to enhance the performance of the training model. The preliminary test set and the final test set data use two known scanners and two unknown scanners. All four scanners are used for STM but only three scanners (Hamamatsu XR, Hamamatsu S360, Aperio CS2) are used for the detection network because the images from Leica GT450 does not contain annotations.

RetinaNet [9] is used for our detection network as the baseline approach in [20] with ImageNet pretrained ResNet 101 backbone. Since the bounding box size of the ground truth is fixed at 50, we set the size of the base anchor to 50 and then gave some variation as shown in Eq. 1. When a fixed anchor size of 50 is used, the results are lower than when various anchor size, which because the detection model is classified with more diverse information through some variations.

$$A = [x \times 2^0, x \times 2^{\frac{1}{3}}, x \times 2^{\frac{2}{3}}], \forall x \in A_{base} \tag{1}$$

Training. Most hyperparameter settings for STM follow the original StarGAN training [2]. The learning rate for both D and G are lowered to 0.00005 because cell images show little variance compared to general images. For the same reason, the total iterations are also lowered to 80 K. After training, the generator is used as STM. The detection network takes an input in which random rotation and random flip augmentation with a style transfer module are applied. Both MF and MLF are classified to learn hard negatives effectively. The model is trained over 120 epochs for approximately a day with a 0.001 learning rate, decaying by $1/10$ on (84600, 112800) steps. $\alpha = 6$ and $p = 0.2$ are chosen experimentally. For each scanner, 40 images are used as training images, 5 as validation images, and 5 as local tests. All training is done on a NVIDIA TITAN RTX GPU.

Inference. All of the patches from WSI are used as input. After detecting both MF and MLF for all patches, the MLF results are discarded, and the remaining results are aggregated. The merging process is held by non-maximum suppression, with the criterion that two objects within 15 pixels are considered the same object.

Table 1. Comparison of F1 score from ablation study

	Hamamatsu XR	Hamamatsu S360	Aperio CS2 (unseen)
baseline	0.7226	0.7576	0.7255
baseline + PS	0.7237	0.7783	0.7643
baseline + PS + STM	0.7532	**0.8307**	0.7586
baseline + PS + a-STM	**0.7895**	0.8293	**0.7816**

4.2 Ablation Study

To determine the effect of our proposed method, an ablation study is conducted for each component (PS, STM, a-STM). In ablation study, we split the MIDOG dataset into local training and test set, and the image from Aperio CS2 scanner is excluded in the training and used in the test phase only to fit the setting of domain generalization. Table 1 shows the effect of each component. Note that solving the imbalance using PS improves the F1 scores from Aperio CS2 because the model is trained similarly for both scanners, making the model more robust. Additionally, using the STM-based model (STM, a-STM) balances the performance between scanners, as expected. However, using the STM-based model shows little degradation on Aperio CS2 scanner, which is the result of patch artifacts described in Sect. 3.2. Therefore, the removal of artifacts in the generated patches prevents the image from losing style details; hence, the results show outstanding performance for all scanners.

Table 2. Comparison of final test set results at the MIDOG Challenge

	Test Set		
	F1	Precision	Recall
reference [20]	0.7108	0.6909	0.7319
ours	**0.7243**	0.6790	0.7761
	F1	Precision	Recall
Hamamatsu XR	**0.8363**	0.8556	0.8177
Leical GT450	0.6448	0.6982	0.5990
3dHistech P1000	0.7650	0.8333	0.7071
Hamamatsu RS	**0.6216**	0.7041	0.5565

5   Conclusion

In this study, we propose a new training scheme using STM and PS to solve the domain generalization problem in mitosis detection. For the preliminary test set, our model using the previous version of the STM achieved an F1 score of 0.7548, ranking 4th in the challenge leaderboard. Table 2 shows the F1 score results on the final test set, suggesting that the use of STM and PS allows the model to function for unseen datasets. Our model achieves a better F1 score of 0.7243 (R: 0.6789, P: 0.7761) than the original reference code. For each scanner, we have obtained a slightly imbalanced score between scanners (0.62 to 0.83). We confirm that the artifact issue explained in Sect. 3.2 is frequently shown in Leical GT450 dataset, which resulted in lower performance for all test images. Furthermore, we have anticipated the score of Hamamatsu RS scanner has dropped because of its feature similarity with Leical GT450. As shown in the ablation study, the results from a-STM which is investigated after the challenge ended on the local dataset suggest that the maintenance of the unique style affects the performance of style transfer by reducing the difference between scanner results. Thus, our complete method using the advanced style transfer module is expected to show greater performance on a real test set.

References

1. Aubreville, M., et al.: Mitosis domain generalization challenge (2021). https://doi.org/10.5281/zenodo.4573978
2. Choi, Y., Choi, M., Kim, M., Ha, J.W., Kim, S., Choo, J.: Stargan: unified generative adversarial networks for multi-domain image-to-image translation. In: Proceedings of the IEEE Conference on Computer Vision and Pattern Recognition, pp. 8789–8797 (2018)
3. Das, D.K., Dutta, P.K.: Efficient automated detection of mitotic cells from breast histological images using deep convolution neutral network with wavelet decomposed patches. Comput. Biol. Med. **104**, 29–42 (2019)
4. Goodfellow, I.J., et al.: Generative adversarial networks (2014)

5. Huang, X., Belongie, S.: Arbitrary style transfer in real-time with adaptive instance normalization. In: Proceedings of the IEEE International Conference on Computer Vision, pp. 1501–1510 (2017)
6. Jin, X., Lan, C., Zeng, W., Chen, Z.: Feature alignment and restoration for domain generalization and adaptation. arXiv preprint arXiv:2006.12009 (2020)
7. Khan, A.M., El-Daly, H., Rajpoot, N.M.: A gamma-gaussian mixture model for detection of mitotic cells in breast cancer histopathology images. In: Proceedings of the 21st International Conference on Pattern Recognition (ICPR2012), pp. 149–152. IEEE (2012)
8. Li, Y., Mercan, E., Knezevitch, S., Elmore, J.G., Shapiro, L.G.: Efficient and accurate mitosis detection-a lightweight RCNN approach. In: ICPRAM, pp. 69–77 (2018)
9. Lin, T.Y., Goyal, P., Girshick, R., He, K., Dollár, P.: Focal loss for dense object detection. In: Proceedings of the IEEE International Conference on Computer Vision, pp. 2980–2988 (2017)
10. Macenko, M., et al.: A method for normalizing histology slides for quantitative analysis. In: 2009 IEEE International Symposium on Biomedical Imaging: From Nano to Macro, pp. 1107–1110. IEEE (2009)
11. Odena, A., Dumoulin, V., Olah, C.: Deconvolution and checkerboard artifacts. Distill **1**(10), e3 (2016)
12. Pourakpour, F., Ghassemian, H.: Automated mitosis detection based on combination of effective textural and morphological features from breast cancer histology slide images. In: 2015 22nd Iranian Conference on Biomedical Engineering (ICBME), pp. 269–274. IEEE (2015)
13. Rao, S.: MITOS-RCNN: a novel approach to mitotic figure detection in breast cancer histopathology images using region based convolutional neural networks. arXiv preprint arXiv:1807.01788 (2018)
14. Reinhard, E., Adhikhmin, M., Gooch, B., Shirley, P.: Color transfer between images. IEEE Comput. Graphics Appl. **21**(5), 34–41 (2001)
15. Somavarapu, N., Ma, C.Y., Kira, Z.: Frustratingly simple domain generalization via image stylization. arXiv preprint arXiv:2006.11207 (2020)
16. Tashk, A., Helfroush, M.S., Danyali, H., Akbarzadeh-Jahromi, M.: Automatic detection of breast cancer mitotic cells based on the combination of textural, statistical and innovative mathematical features. Appl. Math. Model. **39**(20), 6165–6182 (2015)
17. Tellez, D., Balkenhol, M., Karssemeijer, N., Litjens, G., van der Laak, J., Ciompi, F.: H and E stain augmentation improves generalization of convolutional networks for histopathological mitosis detection. In: Medical Imaging 2018: Digital Pathology, vol. 10581, p. 105810Z. International Society for Optics and Photonics (2018)
18. Tzeng, E., Hoffman, J., Darrell, T., Saenko, K.: Simultaneous deep transfer across domains and tasks. In: Proceedings of the IEEE International Conference on Computer Vision, pp. 4068–4076 (2015)
19. Vahadane, A., et al.: Structure-preserved color normalization for histological images. In: 2015 IEEE 12th International Symposium on Biomedical Imaging (ISBI), pp. 1012–1015. IEEE (2015)
20. Wilm, F., Breininger, K., Aubreville, M.: Domain adversarial retinanet as a reference algorithm for the mitosis domain generalization (midog) challenge. arXiv preprint arXiv:2108.11269 (2021)

21. Wu, B., et al.: FF-CNN: an efficient deep neural network for mitosis detection in breast cancer histological images. In: Valdés Hernández, M., González-Castro, V. (eds.) MIUA 2017. CCIS, vol. 723, pp. 249–260. Springer, Cham (2017). https://doi.org/10.1007/978-3-319-60964-5_22
22. Zanjani, F.G., Zinger, S., Bejnordi, B.E., van der Laak, J.A., de With, P.H.: Stain normalization of histopathology images using generative adversarial networks. In: 2018 IEEE 15th International Symposium on Biomedical Imaging (ISBI 2018), pp. 573–577. IEEE (2018)
23. Zhou, K., Loy, C.C., Liu, Z.: Semi-supervised domain generalization with stochastic stylematch. arXiv preprint arXiv:2106.00592 (2021)
24. Zhu, J.Y., Park, T., Isola, P., Efros, A.A.: Unpaired image-to-image translation using cycle-consistent adversarial networks. In: Proceedings of the IEEE International Conference on Computer Vision, pp. 2223–2232 (2017)

Two-Step Domain Adaptation for Mitotic Cell Detection in Histopathology Images

Ramin Nateghi[1]([✉]) and Fattaneh Pourakpour[2]

[1] Electrical and Electronics Engineering Department,
Shiraz University of Technology, Shiraz, Iran
r.nateghi@sutech.ac.ir
[2] Iranian Brain Mapping Lab, National Brain Mapping Laboratory, Tehran, Iran
pourakpour@nbml.ir

Abstract. Mitotic figure count is an important prognostic factor for breast cancer grading. However, the mitotic identification often suffers from the domain variations. We propose a two-step domain-invariant mitosis detection method based on Faster RCNN and a convolutional neural network (CNN). We generate various domain-shifted versions of existing histopathology images using a stain augmentation technique, enabling our method to effectively learn various stain domains and achieve better generalization. The performance of our method is evaluated on the preliminary test and final test sets of the MIDOG-2021 challenge, resulting in F1 score of 68.95% and 67.64% respectively. The experimental results demonstrate that the proposed mitosis detection method can achieve promising performance for domain-shifted histopathology images.

Keywords: Mitosis detection · Breast Cancer · Histopathology images · Faster RCNN

1 Introduction

The number of mitotic figures is one of the critical features in Nottingham grading system [1], which is widely used for breast cancer grading. Manual mitosis cell counting is a time-consuming task in which a pathologist analyzes the selected areas on a tissue. In recent decades, with the advent of whole slide imaging scanners, the entire tissue can be digitized as multiple high-resolution images, encouraging us to develop computerized methods for mitosis detection. The similarity between mitosis and other cells is one of the challenges that hampers accurate detection of mitotic figures. In the recent years, several international competitions have been organized to address mitosis detection challenges: MITOS 2012 [2], AMIDA13 [3], MITOS-ATYPIA-14 [4] and Tumor Proliferation Assessment Challenge TUPAC16 [5]. Recently, several methods have been developed and evaluated on the datasets provided by the aforementioned challenges [6–9]. The main goal of these methods was to identify mitoses within the unseen test samples drawn from the same distribution as training samples.

© Springer Nature Switzerland AG 2022
M. Aubreville et al. (Eds.): MIDOG 2021/MOOD 2021/L2R 2021, LNCS 13166, pp. 32–39, 2022.
https://doi.org/10.1007/978-3-030-97281-3_4

Another major difficulty for histopathological image analysis and especially for mitosis detection is the scanner variability, and stain variations in tissue [10], which can derive from differences in staining conditions and tissue preparation and using various scanners. This variability in the image representation can adversely affect the mitosis detection performance, especially when the training and testing data do not come from the same domain distribution. This problem is well-known as domain shift [11]. Several solution approaches have been proposed in recent literature to cope with domain shift problem [12–21]. Stain normalization is one of the approaches that can be used for domain adaptation, which is often used as pre-processing before training the network [12,13]. The stain normalization methods change the color appearance of a source dataset by using the color characteristics of a specific target image. Although the stain normalization methods reduce the color and intensity variations, they sometimes have an adverse effect on the performance due to not preserving detailed structural information of the cells for all domain shifted cases. Also, the color normalization methods alone cannot solve inter-scanner variability problem [14].

Data augmentation is another popular technique that is used for domain shift adaptation [15–17]. The stain normalization methods aim to equalize the color appearance of the images, while stain augmentation techniques expand the training set by creating new images with new color appearances. It has been shown that stain augmentation techniques can significantly improve the model generalization and they usually outperform stain normalization methods [18]. In recent years, several studies have evaluated the impact of domain shift on model performance. Some recent solutions are based on adversarial neural networks [19–21]. These methods aim to remove the domain information from the model representation. To address the domain shift problem for mitosis detection, the MItosis DOmain Generalization (MIDOG) competition was organized [22] to evaluate the impact of domain shift introduced by using different scanners on mitosis detection performance. In the next sections, we propose two-step mitosis detection based on Faster RCNN [23] and a convolutional neural network. We also used a stain augmentation technique for domain adaptation.

2   Dataset

The dataset used in this study is related to the MIDOG competition that was held as a part of the MICCAI-2021 conference [22]. The MIDOG consists of three training, preliminary test, and final test sets containing 200, 20, and 80 breast cancer histopathology images stained with Hematoxylin and Eosin (H&E). The images of training set were scanned by four different scanners, including the Hamamatsu XR NanoZoomer 2.0, the Hamamatsu S360, the Aperio ScanScope CS2, and the Leica GT450, while the preliminary and final test images were scanned by Hamamatsu XR, Leica GT450, and two scanners unknown for model training (3dHistech P1000, Hamamatsu RS). The mitosis cells were annotated by pathologists within the selected region of interest with an area of approximately $2\,mm^2$. The annotations were only provided for images scanned by three scanners

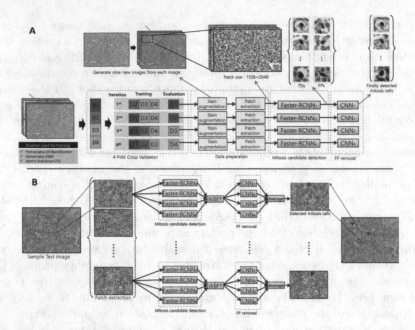

Fig. 1. Block diagram of the proposed mitosis detection method (a) Training phase, (b) Testing phase

in the training dataset, and no annotations were available for the images scanned with Leica GT450. In our experiments, we used the preliminary test and final test sets for performance evaluation.

3   Method

Figure 1 represents the block diagram of our mitosis detection method in training and testing phases. Our method in the training phase consists of four steps: k-fold cross-validation, data preparation, mitosis candidate detection by Faster RCNN models, and false-positive removal by CNN models. For reducing the false-positive detection rate, our method detects mitotic figures in two cascade steps. In the first step, instead of splitting our training dataset into two training and validation subsets, we used the k-fold cross-validation technique (k = 4) as a preventative technique against overfitting. Using this technique, the training dataset containing images captured by Hamamatsu XR NanoZoomer 2.0, Hamamatsu S360, Aperio ScanScope CS2 scanners is randomly divided into four different subsets. We only used the images scanned by these three scanners, since the annotations have been only provided for them.

During the training phase, one of the subsets is considered as a validation set and the remaining as a training set. For domain adaptation and to overcome the domain shifting problem, we used StainTools as a stain augmentation

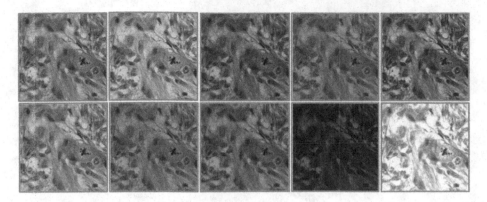

Fig. 2. The stain augmentation examples: (top-left) Original image, (others) Augmented images

technique proposed in [24]. This technique augments an image by decomposing it into the stain matrix S and the concentration matrix C, modifying the concentration matrix, and then recombining it with the stain matrix to get an augmented image. Figure 2 represents some augmented images generated from a sample region of histopathology image. Using the mentioned augmentation technique, we randomly generated nine new histopathology images with shifted stain appearances from each original image. The augmented and original images are then used to train Faster-RCNN and CNN models. This technique helps us expand the training dataset and effectively improve model generalization and performance, especially for unseen domains.

Figure 3 (a) illustrates the domain distribution of training data based on the t-SNE embedding of randomly extracted patches from images of four scanners. As we can see in the figure, the images scanned with Leica GT450 scanner have a quite different distribution than other scanners. This will result in poor performance of the model on unseen images of the Leica GT450 scanner. The performance would be even lower on other unseen distributions. The augmentation technique that we used in our method can effectively cope with this problem. As shown in Fig. 3 (b), the augmented images cover some new domain regions in the t-SNE plot, indicating that the augmentation technique not only enables the model to learn the unseen Leica GT450 scanner distribution but also well-generalizes the model over the other unseen distributions.

Because the images are large in size, in the next step, the images within each subset are split into small patches with the size of 1536×2048 (the padding is done if needed). Then we trained four different Faster RCNN models using the four augmented subsets. For model training, we used a mini-batch size of 4, with a cyclical maximal learning rate of 10^{-4} for 40 epochs by considering binary cross-entropy and smooth L1 losses for classification and regression heads, respectively. The validation loss is also used for the early stopping and checkpoint (with a patience of ten epochs), helping the models to further avoid

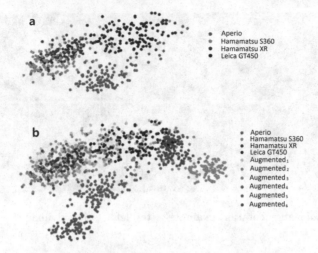

Fig. 3. Illustration of the domain distribution using t-SNE embedding method: (a) Color distribution of four Hamamatsu XR NanoZoomer, Hamamatsu S360, Aperio ScanScope and Leica GT450 scanners (b) Color distribution of augmented images

overfitting. For combining the ensemble results of four trained Faster RCNN models, we used Weighted Boxes Fusion (WBF) method [25]. WBF effectively utilizes confidence scores of all proposed bounding boxes to iteratively constructs the averaged boxes. It has better results than other box ensembling methods like Non-maximum Suppression (NMS).

Having false-positive results is a most challenging problem for mitosis detection. Therefore, the detected mitoses are used to train CNNs to perform finer mitosis detection in the next step. In this step, all of the false-positive and truly detected mitoses at the output of the first step are used to train Efficient-NetB0 networks. Four different networks are trained in the second step using the detected cells within the four subsets. Before training the networks, we extended the cell subsets using the mentioned augmentation technique for domain generalization. For the training, we used a mini-batch size of 256 and trained the models for 200 epochs with a cyclical maximal learning rate of 10^{-4}. To avoid overfitting, the early stopping with a patience of fifty epochs is used during training. The binary cross-entropy loss is also considered to train the networks.

4   Evaluation and Results

We evaluated the performance of the proposed method on training, preliminary and final test sets. Table 1 represents the performance of our method on cross-validation training data across all four folds based on three criteria, including precision, recall, and F1 score. The precision represents the fraction of detected cells that are truly mitosis while recall expresses the fraction of true mitosis cells that are detected. The F1 score is also the harmonic mean of precision

Table 1. Performance of the proposed mitosis detection method on cross-validation training data across all folds

Fold	Faster-RCNNs			Faster-RCNNs + CNNs		
	Precision (%)	Recall (%)	F1 score (%)	Precision (%)	Recall (%)	F1 score (%)
1	53.32	58.15	55.63	85.12	**89.5**	87.25
2	49.56	**60.38**	54.43	83.61	88.23	86.07
3	51.19	58.84	54.74	84.91	86.18	85.54
4	**54.75**	59.27	**56.92**	**86.29**	89.31	**87.77**

Table 2. Performance of our method on the preliminary test set

Method	TP	FP	FN	Precision (%)	Recall (%)	F1 score (%)
Faster-RCNNs	140	389	26	26.46	84.33	40.28
Faster-RCNNs + CNNs	114	66	52	63.33	68.67	65.90
Faster-RCNNs + CNNs + Aug.	121	64	45	65.41	72.89	**68.95**

and recall. The confident threshold is a parameter that can affect performance. The higher threshold value will reduce the false positives but, at the same time, increase the false negative results, resulting in higher precision and lower recall. We have chosen various confident threshold values in our experiments and the optimal thresholds for which the best results (F1 score) are obtained are 0.4 for the first phase and 0.6 for the second phase. The results show that the two-step mitosis detection (Faster-RCNNs + CNNs) can effectively improve the mitosis detection performance. Also, we individually evaluated the performance of the first step mitosis detection results (Faster-RCNNs) on the preliminary test set to better understand the importance of the multi-stage classification in reducing the number of false positive detections.

Table 2 summarizes the performance of our mitosis detection method on the preliminary test set. The results show that the first mitosis detector achieved an F1 score of 40.28% on the preliminary test set, containing some false positives at the output. According to Table 2, the second mitosis classification helped our method to achieve an F1 score of 65.90%. In fact, in the second classification step, the false positives were considerably removed by CNNs and it significantly improved the mitosis detection performance. Our best result on the preliminary test set was obtained using the two-step mitosis detection method and the mentioned augmented technique, resulting in an F1 score of 68.95%. Table 3 also represents the performance of our mitosis detection method on the final test set. The final test set contains images of two new scanners. Therefore, it could be challenging to detect mitotic figures within images of two new scanners. However, because of using augmentation technique, the models could learn new unseen distributions, helping our method generalize on unseen domains.

Table 3. Performance of our method on the final test set

Scanner	TP	FP	FN	Precision (%)	Recall (%)	F1 score (%)
Hamamatsu XR	158	64	45	71.17	**77.83**	**74.35**
Leica GT 450	110	38	87	**74.32**	55.83	63.76
3dHistech P1000	74	28	25	72.54	74.74	73.62
Hamamatsu RS	74	61	50	54.81	59.67	57.13
Overall	**416**	**191**	**207**	**68.53**	**66.77**	**67.64**

5   Conclusion

In this work, we presented a two-step domain-invariant mitosis detection method based on Faster RCNN and CNN models. We used a stain augmentation technique for domain generalization and dataset expansion. Our results demonstrated that stain augmentation and multi-stage classification lead to a marked improvement of domain generalization for mitosis detection.

References

1. Bloom, H., Richardson, W.: Histological grading and prognosis in breast cancer: a study of 1409 cases of which 359 have been followed for 15 years. Br. J. Cancer **11**(3), 1–14 (1957)
2. Roux, L., Racoceanu, D., Lomenie, N., Kulikova, M., Irshad, H., Klossa, J.: Mitosis detection in breast cancer histological images an ICPR 2012 contest. J. Pathol. Inform. **30**(4), 1–7 (2013)
3. Veta, M., Van Diest, P.J., Willems, S.M., Wang, H., Madabhushi, A., Cruz-Roa, A.: Assessment of algorithms for mitosis detection in breast cancer histopathology images. Med. Image Anal. **20**(1), 237–248 (2015)
4. MITOS-ATYPIA-14 Homepage. https://mitos-atypia-14.grand-challenge.org/. Accessed 10 Nov 2021
5. Veta, M., Heng, Y.J., Stathonikos, N., Bejnordi, B.E., Beca, F., Wollmann, T.: Predicting breast tumor proliferation from whole-slide images: the TUPAC16 challenge. Med. Image Anal. **54**(1), 111–121 (2019)
6. Sebai, M., Wang, X., Wang, T.: MaskMitosis: a deep learning framework for fully supervised, weakly supervised, and unsupervised mitosis detection in histopathology images. Med. Biol. Eng. Comput. **58**(7), 1603–1623 (2020). https://doi.org/10.1007/s11517-020-02175-z
7. Mathew, T., Kini, J.R., Rajan, J.: Computational methods for automated mitosis detection in histopathology images: a review. Biocybern. Biomed. Eng. **41**(1), 64–82 (2020)
8. Nateghi, R., Danyali, H., Helfroush, M.S.: A deep learning approach for mitosis detection: application in tumor proliferation prediction from whole slide images. Artif. Intell. Med. **114**, 102048 (2021)
9. Sohail, A., Khan, A., Wahab, N., Zameer, A., Khan, S.: A multi-phase deep CNN based mitosis detection framework for breast cancer histopathological images. Sci. Rep. **11**(1), 1–18 (2021)

10. Balkenhol, M.C., Tellez, D., Vreuls, W., Clahsen, P.C., Pinckaers, H., Ciompi, F.: Deep learning assisted mitotic counting for breast cancer. Nat. Lab. Invest. **99**, 1596–1606 (2019)
11. Aubreville, M., et al.: Quantifying the scanner-induced domain gap in mitosis detection. arXiv preprint, pp. 1–4 (2021)
12. Vahadane, A., et al.: Structure-preserving color normalization and sparse stain separation for histological images. IEEE Trans. Med. Imaging **35**(8), 1962–1971 (2016)
13. Bug, D., et al.: Context-based normalization of histological stains using deep convolutional features. In: Cardoso, M.J., et al. (eds.) DLMIA/ML-CDS -2017. LNCS, vol. 10553, pp. 135–142. Springer, Cham (2017). https://doi.org/10.1007/978-3-319-67558-9_16
14. Leo, P., Lee, G., Shih, N.N., Elliott, R., Feldman, M.D., Madabhushi, A.: Evaluating stability of histomorphometric features across scanner and staining variations: prostate cancer diagnosis from whole slide images. J. Med. Imaging **3**(4), 1–14 (2016)
15. Lafarge, M.W., Pluim, J.P., Eppenhof, K.A., Veta, M.: Learning domain-invariant representations of histological images. Front. Med. **6**(1), 162–174 (2019)
16. Tellez, D., Balkenhol, M., Karssemeijer, N., Litjens, G., van der Laak, J., Ciompi, F.: H and E stain augmentation improves generalization of convolutional networks for histopathological mitosis detection. In: Medical Imaging Digital Pathology International Society for Optics and Photonics, pp. 1–11 (2018)
17. Tellez, D., et al.: Quantifying the effects of data augmentation and stain color normalization in convolutional neural networks for computational pathology. Med. Image Anal. **58**, 1–13 (2019)
18. Otalora, S., Atzori, M., Andrearczyk, V., Khan, A., Mueller, H.: Staining invariant features for improving generalization of deep convolutional neural networks in computational pathology. Front. Bioeng. Biotechnol. **7**(1), 198–211 (2019)
19. Lafarge, M.W., Pluim, J.P.W., Eppenhof, K.A.J., Moeskops, P., Veta, M.: Domain-adversarial neural networks to address the appearance variability of histopathology images. In: Cardoso, M.J., et al. (eds.) DLMIA/ML-CDS -2017. LNCS, vol. 10553, pp. 83–91. Springer, Cham (2017). https://doi.org/10.1007/978-3-319-67558-9_10
20. Tschuchnig, M.E., Oostingh, G.J., Gadermayr, M.: Generative adversarial networks in digital pathology: a survey on trends and future potential. Patterns **1**(6), 1–11 (2020)
21. Stacke, K., Eilertsen, G., Unger, J., Lundstrom, C.: Measuring domain shift for deep learning in histopathology. IEEE J. Biomed. Health Inform. **2**(1), 325–336 (2021)
22. Aubreville, M., et al.: Mitosis domain generalization challenge. Zenodo (2021). https://doi.org/10.5281/zenodo.4573978
23. Ren, S., He, K., Girshick, R., Sun, J.: Faster R-CNN: towards real-time object detection with region proposal networks. IEEE Trans. Pattern Anal. Mach. Intell. **39**(6), 1137–1149 (2017)
24. StainTools Homepage. https://github.com/Peter554/StainTools. Accessed 10 Nov 2021
25. Solovyev, R., Wang, W., Gabruseva, T.: Weighted boxes fusion: ensembling boxes from different object detection models. Image Vis. Comput. **107**(1), 1–12 (2021)

Domain-Specific Cycle-GAN Augmentation Improves Domain Generalizability for Mitosis Detection

Rutger H. J. Fick[✉], Alireza Moshayedi, Gauthier Roy, Jules Dedieu, Stéphanie Petit, and Saima Ben Hadj

Tribun Health, Paris, France
`rfick@tribun.health`

Abstract. As the third-place winning method for the MIDOG mitosis detection challenge, we created a cascade algorithm consisting of a Mask-RCNN detector, followed by a classification ensemble consisting of ResNet50 and DenseNet201 to refine detected mitotic candidates. The MIDOG training data consists of 200 frames originating from four scanners, three of which are annotated for mitotic instances with centroid annotations. Our main algorithmic choices are as follows: first, to enhance the generalizability of our detector and classification networks, we use a state-of-the-art Residual Cycle-GAN to transform each scanner domain to every other scanner domain. During training, we then randomly load, for each image, one of the four domains. In this way, our networks can learn from the fourth non-annotated scanner domain even if we don't have annotations for it. Second, for training the detector network, rather than using centroid-based fixed-size bounding boxes, we create mitosis-specific bounding boxes. We do this by manually annotating a small selection of mitoses, training a Mask-RCNN on this small dataset, and applying it to the rest of the data to obtain full annotations. We trained the follow-up classification ensemble using only the challenge-provided positive and hard-negative examples. On the preliminary and final test set, the algorithm scores an F1 score of 0.7578 and 0.7361, respectively, putting us as the preliminary second-place and final third-place team on the leaderboard.

Keywords: MIDOG Challenge · Mitosis Detection · Instance Segmentation

1 Introduction

Mitosis detection is a highly challenging task in pathology due to the rarity of the events and the highly variable morphological appearance of a cell undergoing mitosis - some being very clear and others highly ambiguous [17]. While several mitosis detection challenges have been organized over the past years (MITOS12 [13], AMIDA13 [14], MITOS14 [11], TUPAC16 [15]), none of them

© Springer Nature Switzerland AG 2022
M. Aubreville et al. (Eds.): MIDOG 2021/MOOD 2021/L2R 2021, LNCS 13166, pp. 40–47, 2022.
https://doi.org/10.1007/978-3-030-97281-3_5

focused on testing the effect of domain shift on the robustness of a mitosis detection method. The MIDOG challenge [1,2] specifically addresses this by providing training data originating from four different scanners but making the unseen test set (partially) consist of images that are not from these same scanners.

1.1 Dataset

Following the challenge description: the MIDOG training subset consists of 200 whole slide images (WSI) from human breast cancer tissue samples stained with routine H&E dye. The samples were digitized with four slide scanning systems: the Hamamatsu XR NanoZoomer 2.0, the Hamamatsu S360, the Aperio ScanScope CS2 and the Leica GT450, resulting in 50 WSI per scanner. For the slides of three scanners, a selected field of interest sized approximately $2\,mm^2$ (equivalent to ten high power fields) was annotated for mitotic figures and hard negative look-alikes. These annotations were collected in a multi-expert blinded set-up, but with the help of computer augmentation, similar to [4]. For the Leica GT450, no annotations were available. The preliminary and final test set consist of four (at the time undisclosed) scanners, only two of which were also part of the training set, namely the Hamamatsu XR, Leica GT450, 3DHistech P1000 and Hamamatsu RS. The preliminary test set consists of only five WSI from each for four test scanners. This preliminary test set was used for evaluating the algorithms prior to submission and publishing preliminary results on a leaderboard. The final test set consists of 20 additional WSI from the same four test scanners. The evaluation through a Docker-based submission system ensured that the participants had no access to the (preliminary) test images during method development.

2 Materials and Methods

We base our method around a classic cascade approach to detect mitotic instances in H&E-stained images. We first use a Mask-RCNN [8] to detect mitotic candidates in an image. These candidates are then extracted as small patches and given to a classifier ensemble of a ResNet50 [7] and DenseNet201 [9]. The predictions are merged via weighted average and the final score is returned.

To improve the generalizability of the method - which is the main purpose of the challenge - we used a Residual Cycle-GAN [3] to transform each image of the training images into all other available domains. In this way, each mitotic annotation is available in all 4 scanner domains. This differs from standard data augmentation (color, hue, brightness, etc.), in that these are not random shifts in appearance for the training process, but specifically towards domains that we *know* are in the testing set. In Fig. 1 we show a 4×4 grid of images of the 4 domains that we transformed to all other domains.

To improve the information present in the data for training a detector, we use Mask-RCNN to create pixel-wise annotations for all annotated mitotic instances. Since we know where all mitoses are, we use Inkscape to manually annotate the first 100 or so, train a pretrained Mask-RCNN model on this small dataset, and apply it specifically around other known mitoses. We use test-time augmentation (8 rotations and flips) and average the predicted masks for each mitosis, resulting in clean masks for most annotations. The remaining "difficult" cases were manually completed, providing us mitosis-specific bounding boxes for all mitotic instances. The average bounding box diagonal in the dataset is 28.8 ± 7.9 pixels, which is consistent with the MITOS12 dataset [10].

Cycle-GAN Domain-Specific Transformations

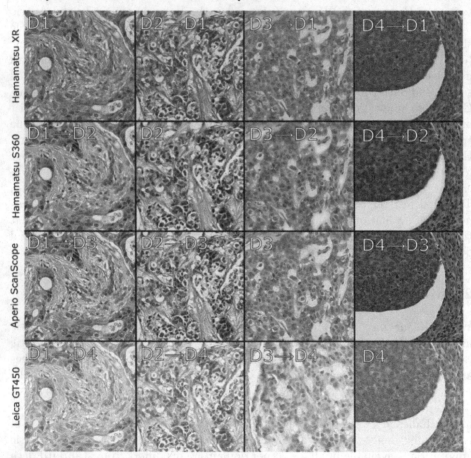

Fig. 1. Residual Cycle-GAN transformed patches. The diagonal are original patches, off-diagonal patches are domain-transformed.

2.1   Domain-Specific Residual Cycle-Gan Augmentation

For the Residual Cycle-GAN [3] we followed the reference model architecture of two sets of generators and discriminators. The Residual Cycle-GAN follows the same principle as a regular one, with the difference that the input image has a direct skip connection with the generated output image. In this way, the generator does not need to reconstruct the image from a set of filter outputs, but only needs to add a "residual", i.e. a color change in the input image so that it resembles a target domain. As it reduces the computational load on the generator, this approach requires fewer data and converges more quickly.

We train six Cycle-GANs to obtain domain transformation functions from all four scanner domains to each other. We train each cycle-GAN for 150000 iterations with generator learning rate $1e-4$, discriminator learning rate $1e-2$, batch size 4, cycle-consistency loss weight 1 and adversarial loss weight 5. We then use the trained generators of these models to create four complete "scanner" copies of the training data, where each copy corresponds to one of four scanners. This means that each "scanner" data set consists of 25% real data and 75% GAN-transformed data, which will be equally sampled from during training. We show an illustration of the 4 data sets in Fig. 1.

2.2   Final Submission Network Training

We split our training data into 45 training slides and 5 validation slides per scanner, ensuring that the validation set had both highly mitotic and non-mitotic slides. The Torchvision implementation of Mask-RCNN with ResNet50 backbone was pretrained on the public COCO2017 dataset, and both the ResNet50 and DenseNet201 classifiers were pretrained on ImageNet. Note that this is the same model architecture as we used for creating the mitosis masks, but now trained on in For Mask-RCNN training, we used a patch size of 3000×3000 pixels and a batch size of 1. We did not train on patches that did not contain any mitoses. We found that using a larger patch size improves the validation performance, and did not improve when adding negative patches. We augmented Mask-RCNN training using skewing, 8 random flips/mirroring, and the domain-specific Cycle-Gan augmentation stated before. We used SGD with a plateau-reduction learning rate scheduler starting at 0.002 and reducing by a factor of 2 if the PR-AUC does not improve after 5 epochs. We warmed up the optimizer during the first epoch and only unfroze the last two convolutional blocks of the Mask-RCNN network. We ran the algorithm until convergence after around 200 epochs.

The classification networks were only trained with the positive and negative instances provided by the challenge organizers - we found that adding hard negatives detected by the detector did not improve leaderboard performance. We used a batch size of 32, and trained for 100 epochs, and kept the model with the best F1 score. We used ADAM with standard parameters and a Cosine annealing learning rate scheduler starting at 2×10^{-5} with a focal loss. For both networks, we only unfroze the backbone after 5 epochs. We used a patch size of 80×80, which we resized to 224×224 to conform with ImageNet pretraining.

We used our GAN-based domain augmentation, together with H&E specific data augmentation [6], with parameters $n = 3$, $m = 7$. The classification head consists of 3 blocks of convolutions with Relu, batch normalization, and dropout set to 0.5, followed by a fully connected layer to the output.

For both the detector and classifiers, many variations of optimization parameters were tried and the model with the best PR-AUC on validation was selected.

2.3   Ablation Study on Instance Segmentation and Domain-Specific GAN Augmentation

The outcome of the challenge indicated that using label enhancement (i.e. adding instance segmentation masks) for mitosis annotations was a winning ingredient for all winning MIDOG approaches. For this reason, we conducted an ablation study on the aforementioned training strategy to understand what aspect of our algorithm most contributed to our success. Note that we perform this study *on the detection algorithm only*, without the subsequent cascade classifiers.

To this end, we started our experiment with a basic Faster-RCNN network [12] with ResNet50 as a backbone, trained using fixed-size bounding boxes of size 50×50 pixels, centered on the mitosis coordinate. We then gradually increased the complexity of our strategy; first by introducing geometric augmentations e.g. rotations, flips, skewing. Then using the exact bounding box obtained from borders of the mask annotations and finally, using a Mask-RCNN with the actual mitosis masks and an offline GAN-based data augmentation method where we transformed the data from each scanner to a different scanner.

3   Results

In Table 1 we show the results of our ablation study to find what worked best for our mitosis detection algorithm. Note that "F1 val" indicates F1 score on Scanner 1, 2 and 3 images, whereas "F1 val S4" is the F1 score on the same validation images but GAN-transformed to look like the scanner 4 domain.

In Table 2 we finally show a summary of our model's scanner-wise performance statistics on the MIDOG challenge test set and our validation set after training. Note that NA means "Not Available" as these scanners were not available in either the test or training set. We discuss our results in the next section.

For reference, our model's aggregate validation PR-AUC was 0.8823 and F1 was 0.8287. On the *preliminary* test set our approach resulted in the second-highest aggregate F1-score of 0.7577, resulting from a 0.7820 precision and a 0.7349 recall.

Table 1. Ablation study to find optimal mitosis detection strategy. From top to bottom, the algorithm becomes increasingly complex. Describing columns from left to right, first, there is experiment ID, which model we use (Fast-RCNN or Mask-RCNN), whether we used fixed-size or adaptive bounding boxes (based on masks), whether we use the mask itself (for Mask-RCNN), use of skew augmentation and use of domain-specific GAN augmentation. Finally, we report F1 score statistics on the validation set (F1 Val) and the F1 score on the GAN-transformed validation set to scanner 4 (F1 val S4).

exp #	model	adapt. bboxes	mask	skew aug	GAN aug	F1 val	F1 val S4
(1)	Faster-RCNN	×	×	×	×	0.824	0.536
(2)	Faster-RCNN	×	×	✓	×	0.835	0.425
(3)	Faster-RCNN	✓	×	✓	×	0.818	0.493[†]
(4)	Faster-RCNN	✓	×	✓	✓	0.823	0.815[*]
(5)	Mask-RCNN	✓	✓	✓	×	0.812	0.705[†]
(6)	Mask-RCNN	✓	✓	✓	✓	0.813	0.812[*]

Table 2. Precision, recall and F1 scores for all scanners available in the MIDOG train and test set. NA indicates "Not Available", as these scanners were either not available in the test or train set. Note that the validation scores for the Leica GT450 scanner have an asterisk, as this scanner was not annotated in the training data, but we used our GAN approach to evaluate the annotated validation set transformed to the Leica GT450 domain.

	Test			Validation		
Scanner	Precision	Recall	F1	Precision	Recall	F1
Hamamatsu XR	0.669	0.572	0.617	0.618	0.871	0.723
Leica GT450	0.693	0.690	0.692	0.798[*]	0.825[*]	0.812[*]
3DHistech P1000	0.851	0.696	0.766	NA	NA	NA
Hamamatsu RS	0.669	0.572	0.617	NA	NA	NA
Hamamatsu S360	NA	NA	NA	0.775	0.968	0.861
Aperio CS2	NA	NA	NA	0.860	0.804	0.831

4 Discussion and Conclusion

From the MIDOG Challenge results the pattern emerged that the first, second, and third place winners (us) all enhanced the mitosis annotations before using some detection algorithm. For this reason, we studied the effect on domain generalizability of adding either instance segmentation for the annotated mitoses or domain-specific GAN augmentation, shown in Table 1. As the MIDOG data does not have a separate test set available to evaluate generalizability for different algorithm variants, we used our GAN domain augmentation to transform our validation set to resemble the non-annotated scanner 4 (Leica GT450). We observe that the "F1 val" score is similar for all experiments regardless of model or augmentation strategy, indicating that for in-training domains there is no significant effect of adding instance masks or domain-augmentation. However, for "F1 val S4" we found that just adding the instance masks for scanners 1–3 already improved generalizability to simulated scanner 4, going from F1 score 0.493 (Exp 3) to 0.705 (Exp 5). Moreover, we see that adding GAN augmentation improves F1 score for F-RCNN, going from F1 score 0.493 (Exp 3) to 0.815 (Exp 4). The same observation is true for Mask-RCNN, going from F1 score 0.705 (Exp 5) to 0.813 (Exp 6). We note that adding the GAN augmentation seems to obviate the benefit of adding masks (F-RCNN versus mask-RCNN), but we chose to submit the Mask-RCNN approach nonetheless. We note, however, that we don't have access to the test set for this ablation study so don't know if our findings on the simulated validation set generalize to the test set.

Finally, we compare the performance of our algorithm between the train, validation, and test set in Table 2. As is expected, our validation scores are always higher than the test scores. Interestingly enough our algorithm generalizes better to an unseen scanner (3DHistech) than a scanner that was actually in the training dataset (Hamamatsu XR), though we note that this scanner also performs worst among the four train scanners in validation. The Leica GT450 scanner, for which we explicitly used our GAN domain augmentation during training, performs second-best in test, suggesting our approach indeed enhanced the model's generalizing properties to this domain.

On the preliminary test set, it was interesting that the MIDOG reference approach [16], which used a RetinaNet with domain adversarial training, was already among the top competitors on the leaderboard. The computational benefit of domain adversarial training over domain-specific GAN augmentation is that it is not necessary to train a cycle-GAN or transform any of the training images. On the other hand, the GAN augmentation can be used for any network architecture without having to choose where to plug in the domain adversarial loss during training - something that the reference approach had to experiment with. It is a subject of future work which of these approaches provides the best domain generalizability.

As for the training of the Residual Cycle-GAN, we note that visually the results illustrated in Fig. 1 seem convincing, but the color transformation is not always completely consistent between different frames. As is typical of GANs, it is hard to know exactly when to stop training, and it is hard to assess how these color variations impact the final mitosis detection performance.

In conclusion, while the winning approaches in the MIDOG challenge were different, it seems that injecting more information into the mitosis detection problem improves the final detection performance. It would be interesting to see how using self-supervised contrastive learning as pretraining [5], instead of ImageNet pretraining, could further improve the mitosis detection performance of any approach.

References

1. Aubreville, M., et al.: MItosis DOmain Generalization challenge (MIDOG). In: 24th International Conference on Medical Image Computing and Computer Assisted Intervention (MICCAI) (2021). https://doi.org/10.5281/zenodo.4573978
2. Aubreville, M., et al.: Quantifying the scanner-induced domain gap in mitosis detection. In: Medical Imaging with Deep Learning (MIDL) (2021)
3. de Bel, T., et al.: Residual cyclegan for robust domain transformation of histopathological tissue slides. Med. Image Anal. **70**, 102004 (2021)
4. Bertram, C.A., et al.: Are pathologist-defined labels reproducible? Comparison of the TUPAC16 mitotic figure dataset with an alternative set of labels. In: Cardoso, J., et al. (eds.) IMIMIC/MIL3ID/LABELS -2020. LNCS, vol. 12446, pp. 204–213. Springer, Cham (2020). https://doi.org/10.1007/978-3-030-61166-8_22
5. Ciga, O., Xu, T., Martel, A.L.: Self supervised contrastive learning for digital histopathology. Mach. Learn. Appl. **7**, 100198 (2021)
6. Faryna, K., van der Laak, J., Litjens, G.: Tailoring automated data augmentation to H&E-stained histopathology. In: Medical Imaging with Deep Learning (2021)
7. He, K., et al.: Deep residual learning for image recognition. In: Proceedings of the IEEE Conference on Computer Vision and Pattern Recognition, pp. 770–778 (2016)
8. He, K., et al.: Mask R-CNN. In: Proceedings of the IEEE ICCV, pp. 2961–2969 (2017)
9. Huang, G., et al.: Densely connected convolutional networks. In: Proceedings of the IEEE Conference on Computer Vision and Pattern Recognition, pp. 4700–4708 (2017)
10. Kausar, T., et al.: SmallMitosis: small size mitotic cells detection in breast histopathology images. IEEE Access **9**, 905–922 (2020)
11. MITOS14 Challenge (2014). https://mitos-atypia-14.grand-challenge.org/
12. Ren, S., et al.: Faster R-CNN: Towards Real-Time Object Detection with Region Proposal Networks (2016). arXiv: 1506.01497 [cs.CV]
13. Roux, L., et al.: Mitosis detection in breast cancer histological images an ICPR 2012 contest. J. Pathol. Inform. **4** (2013)
14. Veta, M., et al.: Assessment of algorithms for mitosis detection in breast cancer histopathology images. Med. Image Anal. **20**(1), 237–248 (2015)
15. Veta, M., et al.: Predicting breast tumor proliferation from whole-slide images: the TUPAC16 challenge. Med. Image Anal. **54**, 111–121 (2019)
16. Wilm, F., Breininger, K., Aubreville, M.: Domain adversarial RetinaNet as a reference algorithm for the MItosis DOmain Generalization (MIDOG) challenge. In: Biomedical Image Registration, Domain Generalisation and Out-of-Distribution Analysis, MICCAI 2021 Challenges L2R, MIDOG and MOOD (2021)
17. Wilm, F., et al.: Influence of inter-annotator variability on automatic mitotic figure assessment. In: Palm, C., Deserno, T.M., Handels, H., Maier, A., Maier-Hein, K., Tolxdorff, T. (eds.) Bildverarbeitung für die Medizin 2021. I, pp. 241–246. Springer, Wiesbaden (2021). https://doi.org/10.1007/978-3-658-33198-6_56

Stain-Robust Mitotic Figure Detection for the Mitosis Domain Generalization Challenge

Mostafa Jahanifar[1]([✉])(iD), Adam Shepard[1](iD), Neda Zamanitajeddin[1](iD),
R. M. Saad Bashir[1], Mohsin Bilal[1], Syed Ali Khurram[2](iD), Fayyaz Minhas[1](iD),
and Nasir Rajpoot[1](iD)

[1] Tissue Image Analytics Centre, Department of Computer Science,
University of Warwick, Coventry, UK
mostafa.jahanifar@warwick.ac.uk
[2] School of Clinical Dentistry, University of Sheffield, Sheffield, UK

Abstract. The detection of mitotic figures from different scanners/sites remains an important topic of research, owing to its potential in assisting clinicians with tumour grading. The MItosis DOmain Generalization (MIDOG) challenge aims to test the robustness of detection models on unseen data from multiple scanners for this task. We present a short summary of the approach employed by the **TIA Centre** team to address this challenge. Our approach is based on a hybrid detection model, where mitotic candidates are segmented on stain normalised images, before being refined by a deep learning classifier. Cross-validation on the training images achieved the F1-score of 0.786 and 0.765 on the preliminary test set, demonstrating the generalizability of our model to unseen data from new scanners.

Keywords: Mitosis detection · MIDOG · Domain generalization · Deep learning

1 Introduction

The detection of mitotic figures is an important task in the analysis of tumour regions [1]. The abundance, or count, of mitotic figures has been shown to be strongly correlated with cell proliferation, which in turn is an important prognostic indicator of tumour behaviour, and thus is a key parameter in several tumour grading systems [1,2]. However, other imposter/mimicker cells are often mistaken for mitotic figures due to their similar appearance/morphology, leading to large inter-rater variability. The introduction of deep learning methods for automated detecting/counting of mitotic figures in histology images offers a potential solution to this challenge.

An additional challenge is the translation of machine learning models into clinical practice (i.e., on whole-slide images or WSIs generated by digital slide scanners), which requires a high degree of robustness to staining and scanner variations.

© Springer Nature Switzerland AG 2022
M. Aubreville et al. (Eds.): MIDOG 2021/MOOD 2021/L2R 2021, LNCS 13166, pp. 48–52, 2022.
https://doi.org/10.1007/978-3-030-97281-3_6

The WSIs can vary in their appearance as a result of differences in the way in which the sample was prepared (e.g. preparation/staining procedures) and scanner acquisition method and particular scanner settings. The result of this variation is a *domain shift* between WSIs collected from different scanners/sites.

The MItosis DOmain Generalization (MIDOG) challenge [3] provides a means of testing different algorithms on cohorts of expertly annotated histology images for mitotic figure detection in the presence of a domain shift. To combat these challenges, we first normalise the stain intensities of all images provided to our model before passing images through our proposed hybrid mitosis detection pipeline. The hybrid analysis pipeline consists of (a) a mitotic candidate segmentation model and (b) refinement by a deep learning (DL) classifier. We generated ground truth (GT) segmentation masks of mitotic figures via a semi-automated DL model [4,5].

The use of a pre-trained DL method for generating GT annotations allows the DL models to exploit important contextual information by treating this detection task as a segmentation task instead.

2 Methodology

2.1 Image Pre-processing

As stain variation is the dominant challenge when analysing histology images from various scanners, in the first step of our proposed pipeline we used Vahadane et al.'s method [6] to normalise the stain intensities of all images in the training set to the target *image 009*, with the help of TIAtoolbox[1] library. Note the same stain matrix acquired from *image 009* is used on-the-fly during the prediction on test images.

2.2 Mitosis Candidate Segmentation

Mitosis Mask Generation. We approach the mitosis candidate detection problem as a segmentation task. However, in order to train a CNN for the segmentation task in a supervised manner, GT masks of the desired objects within the image are required. Since the organizers have only provided approximate bounding box annotations for each mitosis in the released MIDOG dataset, we obtained mitotic instance segmentation masks using NuClick[2] [4,5], a CNN-based interactive segmentation model capable of generating precise segmentation masks for each mitotic figure from a point annotation within the mitotic figure. Therefore, for each annotation point in the dataset, we fed the centre point of the bounding box alongside the patch from the original image into NuClick to generate the individual segmentation mask.

---

[1] https://github.com/TissueImageAnalytics/tiatoolbox.
[2] https://github.com/navidstuv/NuClick.

Segmentation Model. We employed a lightweight segmentation model, called Efficient-UNet [7], for the segmentation task. The Efficient-UNet is a fully convolutional network based on an encoder-decoder design paradigm where the encoder branch is the B0 variant of Efficient-Net [8]. Using this model with pretrained weights from *ImageNet* as a backbone allows the overall model to benefit from transfer-learning, by extracting better feature representations and gaining higher domain generalizability. The Jaccard loss function [9] is robust against the imbalanced population of positive and negative pixels in the segmentation dataset, and thus has been utilised to train the model.

Model Training. In order to train and evaluate the model, we extracted 512×512 patches from the stain-normalised images. There was a large class imbalance in the training dataset, owing to the much fewer patches that contained mitosis (positive patches) in comparison to those without mitosis (negative patches). Since we did not wish to introduce a bias towards predicting empty maps (hence increasing the number of false negatives), we devised an on-the-fly under-sampling approach which guaranteed that similar number of positive and negative patches were sampled at the beginning of each epoch. Here, we used all positive patches in all epochs but randomly sampled the negative patches in each epoch. This way we trained a segmentation model that maintains a high level of precision whilst having a high recall.

Post-processing and Candidate Extraction. At the inference stage of the previous step, each image is tiled with overlap (512×512 patches with 75 pixels overlap) and results for all tiles are aggregated to generate the segmentation prediction map. We then use a sequence of morphological operations and compute the centroid of the connected components to extract candidate mitotic cells from the segmentation map.

2.3 Mitosis Candidates Refinement

In the final step of our method, the mitosis candidates discovered in the previous step were verified using a classifier. Here, we trained an Efficient-Net-B7 [8] classifier to distinguish between mitoses and mimickers. To train the classifier we extracted mitosis and mimicker patches (96×96 pixels) based on the annotations provided by the challenge organizers. Again, to deal with the problem of class imbalance, we incorporated the on-the-fly under-sampling technique.

2.4 Data Augmentation

To make both segmentation and classification networks more robust against the variation seen in histology images, we include the standard data augmentation techniques during the network training phase. The extent and combinations of these augmentation techniques are randomly selected on-the-fly and differ from epoch to epoch.

2.5   Inference

The same pipeline as used for training was applied to each input image for inference. However, in order to benefit from all the models and all the training data, we also included "model ensembling" and "test time augmentation" (TTA) techniques in the inference pipeline. Therefore, during segmentation and classification, predictions from all three models from the cross-validation experiments (in addition to predictions on input image variations by TTA techniques like image flipping and sharpening) are averaged to make more confident and robust final predictions on unseen data.

3   Evaluation and Results

The training set released with the MIDOG challenge contains 150 images with GT annotations. All segmentation and classification models were evaluated in a cross-validation framework as follows: three folds were created based on the images from different scanners (fold 1: images 1–50; fold 2: images 51–100; fold 3: images 101–150). Three experiments were conducted, where the models were trained on two folds and validated on the final fold, such that all images were tested once. Our training scheme simulated the way in which the challenge is tested i.e. the test scanner is not used during the network training.

Many configurations for the segmentation and classification networks were tested, but the ones with minimum segmentation loss and best classification F1-score on the validation set were selected. In Table 1, results of the cross-validation experiments for the segmentation only and the hybrid (segmentation+classification) models are reported separately. The segmentation model alone achieved a F1-score of 0.755 in mitosis detection over all the images in the training set. In comparison, the addition of the classier (mitosis candidate refinement step) increased the F1-score to 0.786. Note that the threshold values and hyper parameters on each step in the proposed pipeline are selected based on the cross-validation experiments.

Table 1. Results of cross-validation experiments on the MIDOG dataset.

	t_{seg}	t_{cls}	Recall	Precision	F1-Score
Segmentation only	0.5	–	0.824	0.696	0.755
Segmentation + Classification	0.4	0.6	0.771	0.801	0.786

Using the proposed hybrid method and the above inference method, we were able to achieve the F1-score of 0.765 on the preliminary test set. The small reduction in F1-score on testing sets, when compared to cross-validation, indicates the robustness and generalizability of the proposed approach.

4 Discussion and Conclusion

In this work, we have presented a new method for the challenge of mitotic figure detection in histology images in the presence of a domain shift. Our proposed method first segments mitotic figures, based on the Efficient-UNet architecture, before passing the results of segmentation on to a DL-based classifier to further differentiate between mitotic figures and hard negatives (mimickers). All images were normalised to a chosen sample image during training, before being normalised on-the-fly during inference. The proposed method achieved a high F1-score of 0.765 when tested on the preliminary test set for the MIDOG challenge.

References

1. Veta, M., et al.: Assessment of algorithms for mitosis detection in breast cancer histopathology images. Med. Image Anal. **20**(1), 237–248 (2015)
2. Aubreville, M., et al.: Deep learning algorithms out-perform veterinary pathologists in detecting the mitotically most active tumor region. Sci. Rep. **10**(1), 1–11 (2020). https://doi.org/10.1038/s41598-020-73246-2
3. Aubreville, M., et al.: Mitosis domain generalization challenge. In: 24th International Conference on Medical Image Computing and Computer Assisted Intervention (MICCAI 2021) (2021). https://doi.org/10.5281/zenodo.4573978
4. Jahanifar, M., Koohbanani, N.A., Rajpoot, N.: Nuclick: from clicks in the nuclei to nuclear boundaries. In: MICCAI 2019 Workshop COMPAY (2019)
5. Koohbanani, N.A., Jahanifar, M., Tajadin, N.Z., Rajpoot, N.: Nuclick: a deep learning framework for interactive segmentation of microscopic images. Med. Image Anal. **65**, 101771 (2020)
6. Vahadane, A., et al.: Structure-preserving color normalization and sparse stain separation for histological images. IEEE Trans. Med. Imaging **35**(8), 1962–1971 (2016)
7. Jahanifar, M., Tajeddin, N.Z., Koohbanani, N.A., Rajpoot, N.M.: Robust interactive semantic segmentation of pathology images with minimal user input. In: 2021 IEEE/CVF International Conference on Computer Vision Workshops (ICCVW), pp. 674–683 (2021)
8. Tan, M., Le, Q.: Efficientnet: rethinking model scaling for convolutional neural networks. In: International Conference on Machine Learning, PMLR, pp. 6105–6114 (2019)
9. Jahanifar, M., Tajeddin, N.Z., Koohbanani, N.A., Gooya, A., Rajpoot, N.: Segmentation of skin lesions and their attributes using multi-scale convolutional neural networks and domain specific augmentations, arXiv preprint arXiv:1809.10243 (2018)

MitoDet: Simple and Robust Mitosis Detection

Jakob Dexl[1,2]([✉]) [iD], Michaela Benz[1] [iD], Volker Bruns[1] [iD], Petr Kuritcyn[1] [iD],
and Thomas Wittenberg[1,2] [iD]

[1] Fraunhofer-Institute for Integrated Circuits IIS, Erlangen, Germany
jakob.dexl@fau.de
[2] University of Erlangen-Nuremberg, Erlangen, Germany

Abstract. Mitotic figure detection is a challenging task in digital pathology that has a direct impact on therapeutic decisions. While automated methods often achieve acceptable results under laboratory conditions, they frequently fail in the clinical deployment phase. This problem can be mainly attributed to a phenomenon called domain shift. An important source of a domain shift is introduced by different microscopes and their camera systems, which noticeably change the colour representation of digitized images. In this method description, we present our submitted algorithm for the Mitosis Domain Generalization Challenge [1], which employs a RetinaNet [5] trained with strong data augmentation and achieves an F1 score of 0.7138 on the preliminary test set.

Keywords: Mitosis detection · Domain generalization · Digital pathology

1 Methods

Motivated by recent data-centric approaches we use a RetinaNet [5] trained with strong data augmentation to enforce prediction consistency.

1.1 Dataset

We use the publicly available Mitosis Domain Generalization Challenge (MIDOG) dataset [1]. The data consists of 200 Whole Slide Images (WSIs) from hematoxylin and eosin (HE) stained breast cancer cases. Furthermore, the dataset can be divided into subsets of 50 images, which were acquired and digitized with four different scanners (Aperio ScanScope CS2, Hamamatsu S360, Hamamatsu XR NanoZoomer 2.0, Leica GT450). For three scanners annotations for mitotic figures and hard negatives (imposters) are provided. The disclosed preliminary and final test sets contain samples of two known scanners and two unknown ones.

This work was supported by the Bavarian Ministry of Economic Affairs, Regional Develop- ment and Energy through the Center for Analytics - Data - Applications (ADA-Center) within "BAYERN DIGITAL II" and by the BMBF (16FMD01K, 16FMD02 and 16FMD03).

© Springer Nature Switzerland AG 2022
M. Aubreville et al. (Eds.): MIDOG 2021/MOOD 2021/L2R 2021, LNCS 13166, pp. 53–57, 2022.
https://doi.org/10.1007/978-3-030-97281-3_7

1.2   Model

Our object detection algorithm consists of a RetinaNet [5] with an EfficientNet B0 [11] backbone. The backbone is initialized with state of the art ImageNet weights, which were trained using RandAugment [2] and Noisy Student [13]. We did not change the feature pyramid and used all five pyramid levels. The network's heads consist of four layers with a channel size of 128. Anchor ratios are set to one while the differential evolution search algorithm introduced by [14] is employed to determine three anchor scales (0.781, 1.435, 1.578) (Fig. 1).

1.3   Domain Generalization Through Augmentation

Our main method to approach domain generalization is data augmentation. Data-driven approaches such as RandAugment [2] have been proven to increase model robustness and have been used in recent state of the art models. Inspired by Trivial Augment [7] a very simple random augmentation strategy is used, where a single augmentation is applied to each image. The augmentations are drawn uniformly from a set of color, noise and special transformations while the augmentation strength is random to some defined degree. The pool of augmentations consists of color jitter, HE [12], fancy PCA, hue, saturation, equalize, random contrast, auto-contrast, contrast limited adaptive histogram equalization (CLAHE), solarize, solarize-add, sharpness, Gaussian blur, posterize, cutout, ISO noise, JPEG compression artefacts, pixel-wise channel shuffle and Gaussian noise. In addition, every image is randomly flipped and RGB channels are randomly shuffled.

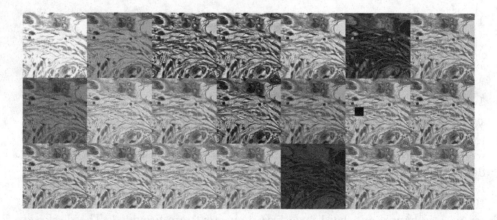

Fig. 1. Used augmentations with different strengths.

1.4   Training and Evaluation

For experimentation, we divide the dataset into five folds with three training, one validation and one test split for each scanner (test splits are added to the train set for submissions). During the training phase, we uniformly sample the images of the train set and randomly select a mitotic figure or an imposter annotation. A patch with a size of 448 pixels is randomly cropped around the selected annotation similar to [6]. The RetinaNet is trained for 100 pseudo epochs with a batch size of 16 using the super-convergence scheme [8]. Adam optimizer with a maximum learning rate of 1e-4 is used. The best models are selected based on the lowest validation loss. After the training phase, we combine the training and validation set and optimize the model's confidence threshold with respect to the best F1 score. During inference, incoming WSIs are tiled into overlapping patches of 448 pixels. All models are trained and tested using an Nvidia GeForce RTX 3060 with 12 GB GPU RAM.

2   Results

For the final submission, we only use labelled data to train a single RetinaNet with the proposed data augmentation strategy. This method achieves an F1 score of 0.7138 on the preliminary test set of the MIDOG challenge.

3   Discussion

Overall, we are able to generalize better across multiple scanner domains with strong data augmentation. The magnitude at which such simple transformations improve generalization at no cost of inference speed is higher than expected. Even models trained with only one scanner reach similar results on our test split, showing only a small performance drop. In the following, we will lay out unsuccessful attempts to improve the quality further. One major issue was the model selection based on the validation loss. The models were not capable of overfitting the data, assumingly due to the sampling and the strong data augmentation, models ended up in an equilibrium mode where performance improvements were wiggling between the different scanners back and forth. Because of that, the representation shift metric proposed by Stacke et al. [10] was tested. It was applied to the three convolutional layers, which flow into the feature pyramid, but was found to not help the model selection process. Another strategy was a dual-stage attempt with a verification net proposed by Li et al. [4]. The network was trained on the predicted patches of the first stage using the same augmentation and in addition a Gradient Reversal Layer [3] to remove even the last bits of scanner dependent information. Unfortunately, this resulted in a performance drop of 12.1% on the preliminary dataset. Finally, the choice of using an EfficientNet originated from the attempt to incorporate the unlabeled data using a self-supervised Student-Teacher learning procedure based on the STAC framework [9]. While increasing the performance on our test split, this resulted in a small performance drop of 1% on the preliminary dataset. One problem was that producing pseudo labels with a high confidence threshold resulted in very few labelled samples while self-training reportedly needs a huge amount of pseudo labelled data to make use of it. A second problem arises with false positive pseudo labels. We used a labelled scanner to check the number of wrong labels incorporated in the pseudo labels and found that for mitotic figures pseudo labels were mainly correct while hard negatives actually included a lot of mitotic figures. This probably led to more confusion than having a positive effect.

References

1. Aubreville, M., et al.: Mitosis domain generalization challenge (2021). https://doi.org/10.5281/zenodo.4573978
2. Cubuk, E.D., Zoph, B., Shlens, J., Le, Q.: RandAugment: practical automated data augmentation with a reduced search space. In: Larochelle, H., Ranzato, M., Hadsell, R., Balcan, M.F., Lin, H. (eds.) Advances in Neural Information Processing Systems, vol. 33, pp. 18613–18624. Curran Associates, Inc. (2020)
3. Ganin, Y., et al.: Domain-adversarial training of neural networks. J. Mach. Learn. Res. **17**(59), 1–35 (2016)
4. Li, C., Wang, X., Liu, W., Latecki, L.J.: DeepMitosis: mitosis detection via deep detection, verification and segmentation networks. Med. Image Anal. **45**, 121–133 (2018). https://doi.org/10.1016/j.media.2017.12.002
5. Lin, T.Y., Goyal, P., Girshick, R., He, K., Dollar, P.: Focal loss for dense object detection. In: Proceedings of the IEEE International Conference on Computer Vision, pp. 2980–2988 (2017)
6. Marzahl, C., et al.: Deep learning-based quantification of pulmonary hemosiderophages in cytology slides. Sci. Rep. **10**(1), 9795 (2020). https://doi.org/10.1038/s41598-020-65958-2
7. Müller, S.G., Hutter, F.: TrivialAugment: tuning-free yet state-of-the-art data augmentation. In: Proceedings of the IEEE/CVF International Conference on Computer Vision, pp. 774–782 (2021)
8. Smith, L.N., Topin, N.: Super-convergence: very fast training of neural networks using large learning rates. In: Artificial Intelligence and Machine Learning for Multi-Domain Operations Applications, vol. 11006, pp. 369–386. SPIE, May 2019. https://doi.org/10.1117/12.2520589
9. Sohn, K., Zhang, Z., Li, C.L., Zhang, H., Lee, C.Y., Pfister, T.: A Simple Semi-Supervised Learning Framework for Object Detection. arXiv:2005.04757 [cs], December 2020
10. Stacke, K., Eilertsen, G., Unger, J., Lundstrom, C.: Measuring domain shift for deep learning in histopathology. IEEE J. Biomed. Health Inform. **25**(2), 325–336 (2021). https://doi.org/10.1109/JBHI.2020.3032060
11. Tan, M., Le, Q.: EfficientNet: rethinking model scaling for convolutional neural networks. In: Proceedings of the 36th International Conference on Machine Learning, pp. 6105–6114. PMLR, May 2019. ISSN 2640-3498
12. Tellez, D., et al.: Whole-slide mitosis detection in H&E breast histology using PHH3 as a reference to train distilled stain-invariant convolutional networks. IEEE Trans. Med. Imaging **37**(9), 2126–2136 (2018). https://doi.org/10.1109/TMI.2018.2820199

13. Xie, Q., Luong, M.T., Hovy, E., Le, Q.V.: Self-training with noisy student improves ImageNet classification. In: Proceedings of the IEEE/CVF Conference on Computer Vision and Pattern Recognition, pp. 10687–10698 (2020)
14. Zlocha, M., Dou, Q., Glocker, B.: Improving RetinaNet for CT lesion detection with dense masks from weak RECIST labels. In: Shen, D., et al. (eds.) MICCAI 2019. LNCS, vol. 11769, pp. 402–410. Springer, Cham (2019). https://doi.org/10. 1007/978-3-030-32226-7_45

Multi-source Domain Adaptation Using Gradient Reversal Layer for Mitotic Cell Detection

Satoshi Kondo(✉)

Muroran Institute of Technology, Muroran, Hokkaido 050-8585, Japan
kondo@mmm.muroran-it.ac.jp

Abstract. In this paper, we propose a multi-source domain adaptation method for mitotic cell detection. Our method is two-step approach. The first step is extraction of candidate regions of mitosis, and the second step is classification of the candidate regions to mitosis or non-mitosis. In the second step, we train a deep neural network model that has two classification tasks, namely mitosis/non-mitosis classification and domain classification. The branch for the domain classification has Gradient Reversal Layer for the domain adaptation. Our method does not use all images in the source domain, but uses the selected images in the domain adaptation phase to reduce the storage size of the source domain data.

Keywords: Mitosis Detection · Domain Adaptation · Gradient Reversal Layer

1 Introduction

Mitosis detection is a key component in tumor prognostication for a range of tumors including breast cancer. Scanning microscopy slides with different scanners leads to a significant visual difference, resulting in a domain shift. This domain shift prevents most deep learning models from generalizing to other scanners, resulting in severely reduced performance.

The scope of the proposed method is to detect mitotic data (cells undergoing cell division) from histopathology images scanned by multiple scanners [2]. The training dataset consists of images scanned by four different scanners, three of which are labeled. The training dataset consists of 200 breast cancer cases in total. The test dataset consists of images scanned by 4 different scanners, two of which are the same scanners in the training dataset.

2 Proposed Method

Our method is two-step approach. The first step is extraction of candidate regions of mitosis. In the second step, we classify the candidate regions to mitosis and non-mitosis. In the following, we explain the details of each step.

© Springer Nature Switzerland AG 2022
M. Aubreville et al. (Eds.): MIDOG 2021/MOOD 2021/L2R 2021, LNCS 13166, pp. 58–61, 2022.
https://doi.org/10.1007/978-3-030-97281-3_8

2.1   Extraction of Candidate Regions

We first transform the input RGB images into Blue Ratio (BR) images. The blue ratio image, which accentuates the nuclear dye, is computed as the ratio of the blue channel and the sum of the other two channels [3]. We extract candidate mitotic regions by binary thresholding of the BR image. The regions are cropped as rectangle patches and these patches are the candidate for mitosis.

2.2   Classification of Candidate Regions

We classify the candidate regions into mitosis or non-mitosis. Our histopathology images are scanned by four different scanners, three of which are labeled. By using the labeled images, we train a deep neural network model which has two classification tasks. The first task is mitosis/non-mitosis classification, which is a binary classification, and the second one is scanner classification, which is a three-class classification. We use ResNet [6] as the base model. In ResNet, we remove the final fully connected layer and append two branches at the end of the network. In the first branch, there are three fully connected layers for mitosis/non-mitosis classification. The second branch is for scanner classification and Gradient Reversal Layer [5] followed by three additional fully connected layers. Figure 1 shows an overview of the network structure in our method.

In the supervised training phase, we use all patches extracted from images of three scanners using labeled source domain data. For each scanner, eighty percent of images are used for training data and the rest images are used for validation data. We use cross entropy loss for each classifier and the final loss is the summation of two losses.

We use images of three scanners in a domain adaptation phase which are labeled and treated as source domain data, and one image from a target domain which is a target image for mitosis detection. A general unsupervised domain adaptation setting is not feasible since we need to keep all source domain data in the domain adaptation stage. To do so, we select patches in the source domain and use those selected patches in the domain adaptation phase. For patch selection, we use a classifier trained using the source domain images. When patches are classified by the classifier, we select patches that have high confidence (high probability) as mitosis or non-mitosis.

In the domain adaptation phase, we extract candidate regions from an image from the target domain. For such patches, however, we do not have labels on the mitosis/non-mitosis. These patches are treated as coming from the fourth scanner (the fourth class in the scanner classifier). We use cross entropy loss for mitosis/non-mitosis classification and scanner classification for the source domain data, and cross entropy loss for scanner classification for the target domain data. The final loss is the summation of those losses.

Upon completion of the domain adaptation phase, patches in the target domain are classified into mitotic or non-mitotic regions.

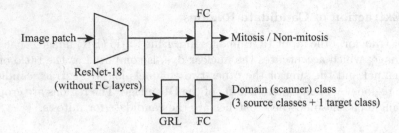

Fig. 1. Network structure of our proposed method.

3   Experimental Conditions and Results

In the extraction of the candidate regions, the threshold is set to the mean value plus 3 times of the standard deviation of the BR image. Regions that are smaller than 2 pixels are discarded, and regions that its width or its height is longer than 50 pixels are also discarded. The size of a patch is 64 × 64 pixels. For example, we selected about 873k patches from 50 images of scanner 1. In these patches, 672 patches were mitotic regions.

We use an 18-layer ResNet as the base network. In the training of the classifier using the source domain data, the learning rate is 1.0×10^{-5}, which is optimized with the Optuna library [1], the batch size is 128, and the number of epochs is 30. The optimizer is Adam [4] and the learning rate is changed with cosine annealing.

For the domain adaptation phase, we select 10k patches from each scanner in the source domain. Out of the 10k patches, there are ten mitotic regions.

In the domain adaptation phase, the learning rate is 1.7×10^{-6}, which is optimized with the Optuna library using the source domain data, the batch size is 128, the number of epochs is 5. We use Adam optimizer and do not change the learning rate.

As shown in the preliminary testing phase using 20 images, the precision, recall and F1 scores of our method are 0.72, 0.74, and 0.73, respectively.

4   Conclusion

We proposed a multi-source domain adaptation method for mitotic cell detection. Our method does not use all patches in the source domain, but uses the selected patches in the domain adaptation phase to reduce the source domain data.

The future work is to improve the method to extract the candidate regions and to select the patches in the source domains.

References

1. Akiba, T., Sano, S., Yanase, T., Ohta, T., Koyama, M.: Optuna: a next-generation hyperparameter optimization framework. In: Proceedings of the 25th ACM SIGKDD International Conference on Knowledge Discovery & Data Mining, pp. 2623–2631 (2019)
2. Aubreville, M., et al.: Mitosis domain generalization challenge (2020). https://doi.org/10.5281/zenodo.4573978
3. Chang, H., Loss, L.A., Parvin, B.: Nuclear segmentation in H&E sections via multi-reference graph cut (MRGC). In: International Symposium Biomedical Imaging (2012)
4. Kingma, D.P., Ba, J.: Adam: a method for stochastic optimization. In: Proceedings of the 3rd International Conference for Learning Representations (ICLR) (2015)
5. Ganin, Y., et al.: Domain-adversarial training of neural networks. J. Mach. Learn. Res. 17(1), 1–35 (2016)
6. He, K., Zhang, X., Ren, S., Sun, J.: Deep residual learning for image recognition. In: Proceedings of the IEEE Conference on Computer Vision and Pattern Recognition, pp. 770–778 (2016)

Rotation Invariance and Extensive Data Augmentation: A Strategy for the MItosis DOmain Generalization (MIDOG) Challenge

Maxime W. Lafarge[(✉)] and Viktor H. Koelzer

Department of Pathology and Molecular Pathology, University Hospital and
University of Zurich, Zurich, Switzerland
`maxime.lafarge@usz.ch`

Abstract. Automated detection of mitotic figures in histopathology images is a challenging task: here, we present the different steps that describe the strategy we applied to participate in the MIDOG 2021 competition. The purpose of the competition was to evaluate the generalization of solutions to images acquired with unseen target scanners (hidden for the participants) under the constraint of using training data from a limited set of four independent source scanners. Given this goal and constraints, we joined the challenge by proposing a straight-forward solution based on a combination of state-of-the-art deep learning methods with the aim of yielding robustness to possible scanner-related distributional shifts at inference time. Our solution combines methods that were previously shown to be efficient for mitosis detection: hard negative mining, extensive data augmentation, rotation-invariant convolutional networks.

We trained five models with different splits of the provided dataset. The subsequent classifiers produced F_1-score with a mean and standard deviation of 0.747 ± 0.032 on the test splits. The resulting ensemble constitutes our candidate algorithm: its automated evaluation on the preliminary test set of the challenge returned a F_1-score of 0.6828.

Keywords: Computational Pathology · Mitosis Detection · Rotation Equivariance/Invariance

1 Dataset Preparation

The organizers of MIDOG 2021 [1] provided annotated images from 150 cases (50 cases each from 3 different source scanners). 50 images from a fourth scanner were provided but we chose not to use them in order to present a solution based solely on a supervised learning framework, thus leaving room for improvements for future work.

We created five folds of three splits such that we were able to train and validate multiple models with varying data distributions. For each fold, we partitioned cases in splits with the following distribution: training (80%), validation

© Springer Nature Switzerland AG 2022
M. Aubreville et al. (Eds.): MIDOG 2021/MOOD 2021/L2R 2021, LNCS 13166, pp. 62–67, 2022.
https://doi.org/10.1007/978-3-030-97281-3_9

(10%) and test (10%), such that the distribution of scanners was identical within each split. With this partition we intended to use as much available source data as possible for training while keeping a small proportion for internal validation and model selection.

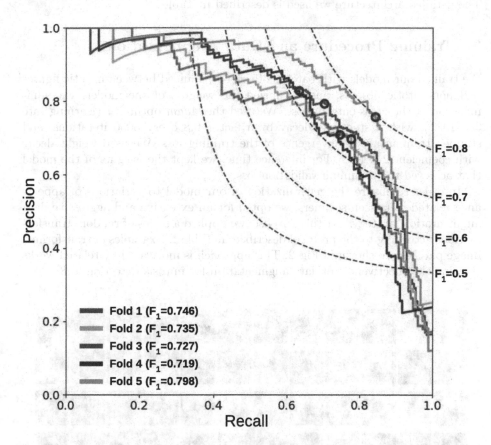

Fig. 1. Precision-Recall analysis of five models trained and evaluated on the different test sets for each fold of the dataset. Dark blue circles show the performances achieved by the models using the operating points that maximized the F_1-score on the validation sets.

2   Model Architecture

We modeled the conditional likelihood of the mitosis class given an input image patch of size 77×77 at magnification $40\times$ using convolutional neural networks (CNNs). Motivated by the benefits of rotation invariance property of deep learning models for computational pathology tasks [2,5,7,9], we used roto-translation equivariant convolutional layers with a 8-fold discretization of the orientation axis [7]. As this structure guarantees the roto-translation equivariance of the

internal activations and invariance of the output of the models with respect to the orientation of the input, rotation augmentation at training and inference time becomes an unnecessary step. Furthermore, this gained invariance property prevents learning possible biases related to the orientation of the images.
The detailed architecture we used is described in Table 1.

3   Training Procedure and Data Augmentation

We trained our models with batches of size 64 balanced between mitotic figures and non-mitotic objects, and optimized the weights of the models via minimization of the cross-entropy loss. We used the *Adam* optimizer (learning rate 3×10^{-4}), with a step-wise decay by a factor 0.8 every 5000 iterations, and stopped training after convergence of the training loss. We used weight decay with coefficient 2×10^{-4}. For inference time, we kept the weights of the model that achieved the minimum validation loss.

In order to ensure the generalization of our model to variations of appearance related to unseen scanners, we opted for an extensive and aggressive data augmentation strategy. For this purpose, we applied a series of random transformations according to the protocol described in Table 2. Examples of transformed image patches are shown in Fig. 2. This approach is motivated by related works showing the effectiveness of data augmentation for mitosis detection [6,8].

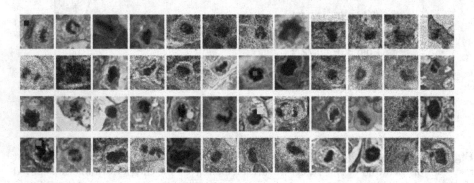

Fig. 2. Example of mitosis-centered image patches transformed according to our random data augmentation protocol.

Generating training batches via random sampling of non-mitotic image patches is known to be a suboptimal approach for mitosis detection as models are less exposed to challenging non-mitotic objects during training [3]. Therefore, to encourage the model to discriminate challenging non-mitotic objects, for each fold, we sequentially resampled the dataset by removing easy classified patches using a protocol derived from [3] using first versions of the models trained via random sampling of the training sets.

Table 1. Architecture of the CNN used in this work. Shape of output tensors are written with the following format: (*Orientations*×)*Channels*(×*Height*× *Width*). Shape of operator tensors are written with the following format: (*Orientations*×)*Out.Ch.*× *In.Ch.*×*Ker.Height*×*Ker.Width*. * indicates that the operation is followed by a *Batch Normalization* layer and a leaky *ReLU* non-linearity (coefficient 0.01).

Layer	Operator Shape	Output Shape
Input	–	$3\times77\times77$
Lifting Convolution *	$16\times3\times4\times4$	$8\times16\times74\times74$
Max Pooling	2×2	$8\times16\times37\times37$
SE(2,8)-Convolution *	$8\times16\times16\times4\times4$	$8\times16\times34\times34$
Max Pooling	2×2	$8\times16\times17\times17$
SE(2,8)-Convolution *	$8\times16\times16\times4\times4$	$8\times16\times14\times14$
Max Pooling	2×2	$8\times16\times7\times7$
SE(2,8)-Convolution *	$8\times16\times16\times4\times4$	$8\times16\times4\times4$
SE(2,8)-Convolution *	$8\times32\times16\times4\times4$	$8\times32\times1\times1$
Maximum Projection	–	32
Fully Connected *	64×32	64
Fully Connected + Sigmoid	1×64	1

Table 2. Data augmentation protocol: for each input image patch, we scanned the following list of transformations and applied it with a given probability, after random sampling of a set of coefficients.

Transformation	Coefficients	Probability
Transposition	–	50%
Color Shift	$c_{r,g,b} \sim U\,[-13, 13]$	50%
Gamma Correction	$\gamma_{r,g,b} \sim U\,[0.9, 1.5]$	50%
Hue Rotation	$h \sim U\,[0, 1]$	50%
Spatial Shift	$\Delta_{x,y} \sim U\,[-12\text{px}, 12\text{px}]$	100%
Spatial Scale	$\alpha \sim U\,[-13\%, 13\%]$	50%
Additive Gaussian Noise	$c_{x,y,c} \sim \mathcal{N}(0, 50)$	50%
Cutout [4] (random color/size s)	$s \sim U\,[8\text{px}, 16\text{px}]$	50%

4   Inference Time

At inference time, the fully convolutional structure of our models enables their dense application on large test images which produces probability maps. Candidate mitotic figures are identified as local maxima after applying non-maxima suppression within a radius of 30px. Our models are then turned into binary classifiers by setting a cutoff threshold (operating point) that is selected such that the F_1-scores on the validation sets were maximized. We applied this procedure to generate a classifier for each fold, and then gathered the 5 models to form an ensemble. The performance of these classifiers on the source test sets are reported in Fig. 1. For new test images the detections of each classifier are considered as votes for candidate mitoses and we filter out detections that get less than 2 votes.

5   Conclusions and Discussion

We proposed a straight-forward approach combining multiple state-of-the-art solutions to tackle the generalization problem for scanner-related distributional shifts in the context of the MIDOG2021 competition. We report a lower performance of our solution on the preliminary test set provided by the organizers compared to the performances we obtained on the source test sets, suggesting that the generalization of our model is limited to some extent. We hope that our methodology can be considered as a baseline, that could potentially be improved using additional training components for domain generalization. In future work, we will aim at investigating the reasons of the generalization limitations of the presented method.

References

1. Aubreville, M., et al.: Mitosis domain generalization challenge. Zenodo (2021). https://doi.org/10.5281/zenodo.4573978
2. Bekkers, E.J., Lafarge, M.W., Veta, M., Eppenhof, K.A.J., Pluim, J.P.W., Duits, R.: Roto-translation covariant convolutional networks for medical image analysis. In: Frangi, A.F., Schnabel, J.A., Davatzikos, C., Alberola-López, C., Fichtinger, G. (eds.) MICCAI 2018. LNCS, vol. 11070, pp. 440–448. Springer, Cham (2018). https://doi.org/10.1007/978-3-030-00928-1_50
3. Cireşan, D.C., Giusti, A., Gambardella, L.M., Schmidhuber, J.: Mitosis detection in breast cancer histology images with deep neural networks. In: Proceedings of the International Conference on Medical Image Computing and Computer-Assisted Intervention (MICCAI), pp. 411–418 (2013)
4. DeVries, T., Taylor, G.W.: Improved regularization of convolutional neural networks with cutout. arXiv preprint arXiv:1708.04552 (2017)
5. Graham, S., Epstein, D., Rajpoot, N.: Dense steerable filter CNNs for exploiting rotational symmetry in histology images. IEEE Trans. Med. Imaging **39**, 4124–4136 (2020)
6. Lafarge, M., Pluim, J., Eppenhof, K., Veta, M.: Learning domain-invariant representations of histological images. Front. Med. **6**, 162 (2019)
7. Lafarge, M.W., Bekkers, E.J., Pluim, J.P., Duits, R., Veta, M.: Roto-translation equivariant convolutional networks: application to histopathology image analysis. Med. Image Anal. **68**, 101849 (2021)

8. Tellez, D., Balkenhol, M., Karssemeijer, N., Litjens, G., van der Laak, J., Ciompi, F.: H and E stain augmentation improves generalization of convolutional networks for histopathological mitosis detection. In: Proceedings of SPIE Medical Imaging, p. 105810Z (2018)
9. Veeling, B.S., Linmans, J., Winkens, J., Cohen, T., Welling, M.: Rotation equivariant CNNs for digital pathology. In: Proceedings of the International Conference on Medical Image Computing and Computer-Assisted Intervention (MICCAI), pp. 210–218 (2018)

Detecting Mitosis Against Domain Shift Using a Fused Detector and Deep Ensemble Classification Model for MIDOG Challenge

Jingtang Liang[1], Cheng Wang[1], Yujie Cheng[2], Zheng Wang[2], Fang Wang[2], Liyu Huang[1], Zhibin Yu[2], and Yubo Wang[1(✉)]

[1] School of Life Science and Technology, Xidian University, Xi'an, Shaanxi, China
huangly@mail.xidian.edu.cn, ybwang@xidian.edu.cn
[2] College of Electrical Engineering, Ocean University of China,
Qingdao, Shandong, China
yuzhibin@ouc.edu.cn

Abstract. Deep learning based mitotic figure detection methods have been utilized to automatically locate the cell in mitosis using hematoxylin & eosin (H&E) stained images. However, the model performance deteriorates due to the large variation of color tone and intensity in H&E images. In this work, we proposed a two stage mitotic figure detection framework by fusing a detector and a deep ensemble classification model. To alleviate the impact of color variation in H&E images, we utilize both stain normalization and data augmentation, leading model to learn color irrelevant features. The proposed model obtains an F1 score of 0.7550 on the preliminary testing set and 0.7069 on the final testing set.

Keywords: Mitosis · Domain shift · data augmentation · deep ensemble model

1 Introduction

Tumor proliferation obtained form hematoxylin & eosin stained (H&E) histopathological images provides valuable information regarding the patient prognosis and treatment planning, especially in breast cancers [6]. Mitotic activity of tumor cells observed in high power field view is an epiphenomenon of the cell proliferation, is therefore selected to quantify the tumor proliferation and has been shown to associated with the patients' prognosis [3].

The large variability observed in H&E stained pathological images still impeded the application of automatic mitosis detection in clinical settings. Despite the source of variation in the H&E images, it mostly manifests as large variation in color tone among different H&E images. Hence, earlier attempts

Supported by the Fundamental Research Funds for the Central Universities, Xidian University, under Grant No. XJS201213.

primarily focused on unifying the color space by utilizing color normalization techniques [2].

In this work, we presented our approach to the MIDOG challenge [1]. Inspired by earlier works [5], we constructed a two stages mitosis detection model by using the detectorRS [4] as the base model to coarsely identify the mitosis figure in the images. The results of detector model is later refined by a deep ensemble classification model to illuminate false positives and improve the overall performance. To address the domain shift problem, we employed both stain normalization and data augmentation focusing on inducing color variation. Our results suggested two-stages model equipped with both stain normalization and data augmentation can be an potential solution to address the domain shift in detecting mitosis figures in H&E images.

2   Data-Set

The data-set was provided by the MIDOG challenge [1]. In brief, all images were obtained from human breast cancer tissue samples after routine Hematoxylin & Eosin staining. The Training set consists of 200 H&E images obtained from four different scanners, including Hamamatsu XR nanozoomer 2.0, Hamamatsu S360 (0.5 NA), Aperio ScanScope CS2, and Leica GT450. Each scanner provided 50 H&E images. Mitotic figures were annotated for the first three scanners. In total, annotation of 1721 and mitotic figures and 2714 non-mitotic figures (hard negative cases) were provided. To train our model, we randomly selected 5 images from each scanner with annotation as the validation set. The rest of training images were used to optimize the model.

The preliminary test set released by the MIDOG challenge consisted of 20 WSIs from four scanners, in which two scanners were part of the training set and the remaining two scanners are unknown.

3   Proposed Model

The proposed model is shown in Fig. 1. Our whole H&E image processing pipeline consisted of five steps. Firstly, we cropped the original training images into patches of the size 512 × 512 pixels, centered at the ground truth mitotic figures and hard negative cases. For each annotated cases, we randomly shifted the center of each patch within the range of ±205 pixels. Then, a detectorRS model [4] was trained to identify the location of mitotic figures using a bounding box with a size of 50 × 50 pixels. In the training phase, all training images were normalized with respect to the first images of the first scanner (*001.tiff*) by using Macenko stain normalization. Then, we augmented the training patches by using random rotation, elastic deformation, scaling, Gaussian blur and a brightness and contrast enhancement. The detector was trained by using SGD with a learning rate of 0.02 for 12 epochs. Once the detector was trained, we employed the trained model to the whole training images to identify all suspected mitotic figures. To be noted, the model was trained using patches containing annotations of ground truth mitotic and hard negative cases, whereas the trained model

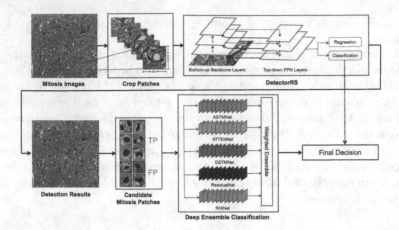

Fig. 1. Pipeline of the proposed mitotic figure detection framework

scanned through the training images can produce many previously un-annotated false mitotic figures. This observation also motivated us to employed a second stage classifier to refine the results produced by the detector.

The overall structure of deep ensemble model consists of five convolution networks, adopted from [5]. The input to the deep ensemble model was the suspected mitotic figures found by the trained detector on the training images with a classification threshold of 0.3. The positive cases for training the classification model consisted of all samples with ground truth mitotic figures. The negative cases were the false positive cases identified by the detector on the whole training images and the hard negative cases. Training samples for the classification model were construed by shifting the center of obtained patches from detector within the range of $[-5, 5]$ pixels. To balance the training samples between positive and negative cases, we adjusted the number of times of applying offset to balance the number of cases in positive and negative classes. Finally, the samples were resized to 120×120 and fed to the deep ensemble model. To overcome the domain shift caused by different scanner, we heavily utilized online augmentation methods that can induce color variation to increase the diversity of the training samples. The augmentation employed were horizontal and vertical flipping, random clipping and color jitter augmentation with luminance, contrast, hue and saturation disturbance intensity. Each individual model was optimized using AdamW with a learning rate of 2×10^{-4} and was trained for 100 epochs. The optimal weights for each individual convolution network was selected based on their performance on the validation set. The output of the ensemble model was the weighted sum of soft-max score produced by each convolution networks. The final decision of the proposed two-stage mitotic figure detection was obtained by combing the classification score obtained from both detector and deep ensemble model as,

$$S_{final} = \alpha * S_{DE} + (1 - \alpha) * S_{Dect} \tag{1}$$

where $\alpha \in [0,1]$ is the weights to balance the decision made by the detector and the deep ensemble model and optimized on the validation set, S_{final} is the final score to produce the final decision, S_{DE} and S_{Dect} are the classification score for the deep ensemble modular and detection modular, respectively.

4   Results

We first tested the performance of detection modular on the validation set. The results of F1 score, precision and recall were given in Table 1. It can be observed that detector alone was able to retrieve almost 80% of mitosis figures. In the meantime, it also produced many false positives resulted in a inferior precision score and a significantly degraded F1-score. The ability to refine the results obtained from the detector by the ensemble classification model weighted by different α was shown in Fig. 2. It can be observed that by varying the value of $\alpha \in [0,1]$, the optimal performance on F1 score was found when $\alpha = 0.9$. The obtained model obtained a F1 score of 0.7550 on the preliminary testing set and 0.7069 on the final testing set.

Table 1. Model performance on validation set, preliminary testing set and final testing set

	Validation Set	Preliminary Testing Set	Final Testing Set
F1-Score	0.7128	0.7550	0.7069
Precision	0.7270	0.7238	0.7279
Recall	0.6993	0.7892	0.6870

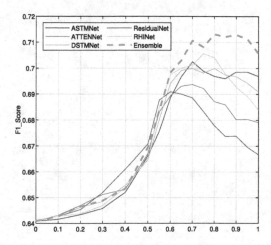

Fig. 2. Performance of individual and ensemble classification model on the validation set

5 Conclusion

In conclusion, we presented a fused detector and deep ensemble classification model with image preprocessed by stain normalization and heavy data augmentation to address the domain shift problem for mitosis figure detection. Experiment results showed that the fused model performs reasonably well on the preliminary testing set released by the MIDOG challenge.

References

1. Aubreville, M., et al.: MItosis DOmain generalization challenge. In: 24th International Conference on Medical Image Computing and Computer Assisted Intervention (MICCAI 2021), pp. 1–15 (2021). https://doi.org/10.5281/zenodo.4573978
2. Macenko, M., et al.: A method for normalizing histology slides for quantitative analysis. In: 2009 IEEE International Symposium on Biomedical Imaging: From Nano to Macro, pp. 1107–1110. IEEE (2009)
3. Medri, L., et al.: Prognostic relevance of mitotic activity in patients with node-negative breast cancer. Mod. Pathol. **16**(11), 1067–1075 (2003)
4. Qiao, S., Chen, L.C., Yuille, A.: DetectoRS: detecting objects with recursive feature pyramid and switchable atrous convolution (2020). http://arxiv.org/abs/2006.02334
5. Sohail, A., Khan, A., Nisar, H., Tabassum, S., Zameer, A.: Mitotic nuclei analysis in breast cancer histopathology images using deep ensemble classifier. Med. Image Anal. **72**, 102121 (2021). https://doi.org/10.1016/j.media.2021.102121
6. Van Diest, P.J., Van Der Wall, E., Baak, J.P.A.: Prognostic value of proliferation in invasive breast cancer: a review. J. Clin. Pathol. **57**(7), 675–681 (2004). https://doi.org/10.1136/jcp.2003.010777

Domain Adaptive Cascade R-CNN for MItosis DOmain Generalization (MIDOG) Challenge

Xi Long[1], Ying Cheng[1], Xiao Mu[1], Lian Liu[2], and Jingxin Liu[1(✉)]

[1] Histo Pathology Diagnostic Center, Shanghai, China
[2] Department of Electrical and Computer Engineering, Technical University of Munich, Munich, Germany

Abstract. We present a summary of domain adaptive cascade R-CNN method for mitosis detection of digital histopathology images. By comprehensive data augmentation and adapting existing popular detection architecture, our proposed method has achieved an F1 score of 0.7500 on the preliminary test set in MItosis DOmain Generalization (MIDOG) Challenge at MICCAI 2021.

Keywords: Mitosis detection · Histopathology · Domain Adaptation

1 Introduction

Mitotic count (MC) is a common and critical marker of breast cancer prognosis [4]. Manually marking mitotic cells in Hematoxylin and Eosin (H&E) stained histopathology images is obviously time-consuming and subjective. With the dramatic improvements in computer vision and digital pathology, researchers proposed to automate this process in pathology laboratories. A number of mitosis detection competitions have been held, e.g., the ICPR MITOS-2012 challenge [9], the ICPR MITOS-ATYPIA-2014 challenge [8], and the MICCAI-TUPAC16 challenge [12]. Thus, numerous works have been proposed, and achieved remarkable success in the field of mitosis detection [2,10].

However, deep learning based detection models may have poor generalization capability to unseen datasets due to the domain shift. Such problem is commonly observed in digital histopathology image analysis, caused by tissue preparation and image acquisition. The MItosis DOmain Generalization (MIDOG) challenge [1], hosted as a satellite event of the 24th International Conference at Medical Image Computing and Computer Assisted Intervention (MICCAI) 2021, addresses this topic in the form of assessing MC on a multiscanner dataset. In this abstract, we propose a method with domain augmentation and Domain Adaptive Cascade R-CNN (DAC R-CNN) for mitosis detection to achieve robust detection performance for varieties of images.

© Springer Nature Switzerland AG 2022
M. Aubreville et al. (Eds.): MIDOG 2021/MOOD 2021/L2R 2021, LNCS 13166, pp. 73–76, 2022.
https://doi.org/10.1007/978-3-030-97281-3_11

2 Materials

The MIDOG training set consists of 200 image tiles from Whole Slide Images
(WSIs) of human breast cancer tissues with H&E dye. The image tiles were
digitized with four slide scanners: Hamamatsu XR nanozoomer 2.0, Hamamatsu
S360 (0.5 NA), Aperio ScanScope CS2, and Leica GT450, resulting in 50 image
tiles per scanner. From each image tile, a trained pathologist selected an area
of $2\,mm^2$ corresponding to approximately 10 high power fields. Annotations are
provided for the first three scanners.

3 Methodology

3.1 Stain Color Domain Augmentation

In MIDOG challenge, the test set contains images scanned by unknown slide
scanners. Previous stain normalization methods transfer different pathology
images into one target stain color style, which may not improve the robustness
of detection model when dealing with unseen stain color appearance. Therefore,
in addition to traditional image augmentation methods, we propose stain color
domain augmentation to generate training images with a wider range of stain
color appearances, making our model more robust to unseen data. To be specific, we build on previous stain normalization methods by adding randomness
in selecting normalization methods and target color styles.

   The proposed stain color domain augmentation method involves two types
of stain normalization methods: *Reinhard* [7] and *Vahadane* [11]. Each method
will be executed with a given probability. *Reinhard* transfers color based on target mean and variance, while *Vahadane* transfers color according to the target
color appearance matrix. We obtain an initial range for target mean, variance,
and each element of the color appearance matrix using the whole training set,
respectively. Target values are randomly selected from those ranges during augmentation, making it possible to generate images with very different color styles.
We will gradually enlarge those ranges to create new training samples to feed
the network until detection performance degrades to a limit. In this way, we
expect the trained network to achieve robust detection performance for varieties
of images.

3.2 Domain Adaptive Cascade R-CNN

We propose a Cascade R-CNN based domain adaptation model for mitosis detection [3], referred to as Domain Adaptive Cascade R-CNN (DAC R-CNN) (See
Fig. 1). The backbone network of DAC R-CNN is pre-trained ResNet-50, and
three cascaded detection heads are utilized for high quality detector. Inspired
by our previous work [5], we employ an image-level adaptation component to
address overall differences between different image domains like image color and
style using PatchGAN [6], through which we aim to obtain similar feature maps
from input image and reference image.

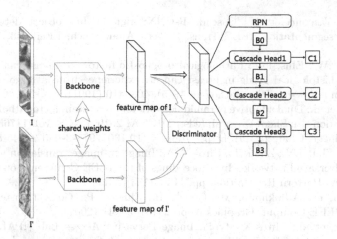

Fig. 1. Domain Adaptive Cascade R-CNN architecture. "I" is input image, "I'" reference image for domain adaptation, "Backbone" backbone network for feature map extraction, "Cascade Head" detection head, "B" bounding box, and "C" classification. "Discriminator" is a convolutional PatchGAN classifier for distinguishing input image from reference image, which should be removed in inference phase.

Specifically, a parallel reference branch is added to the network architecture with a discriminator following feature maps of both the input image and the reference image. Two branches share the same backbone, which serves as the generator. The discriminator distinguishes the input image from the reference image. The discriminator is a convolutional PatchGAN classifier that operates on image patches. One advantage of PatchGAN is that it can be applied to images with arbitrary sizes. Note that labels of reference images are not needed. By generating similar features, domain adaptation seeks to achieve comparable performance for unlabelled data.

4   Results

Our proposed method produced an F1 score of 0.7500 with a 0.7792 precision and a 0.7229 recall on the preliminary test set in MItosis DOmain Generalization (MIDOG) Challenge at MICCAI 2021.

References

1. Aubreville, M., et al.: Mitosis domain generalization challenge (2021). https://doi.org/10.5281/zenodo.4573978
2. Bertram, C.A., Aubreville, M., Marzahl, C., Maier, A., Klopfleisch, R.: A large-scale dataset for mitotic figure assessment on whole slide images of canine cutaneous mast cell tumor. Sci. Data **6**(1), 1–9 (2019)

3. Cai, Z., Vasconcelos, N.: Cascade R-CNN: high quality object detection and instance segmentation. IEEE Trans. Pattern Anal. Mach. Intell. **43**, 1483–1498 (2019)
4. Elston, C.W., Ellis, I.O.: Pathological prognostic factors in breast cancer. I. The value of histological grade in breast cancer: experience from a large study with long-term follow-up. Histopathology **19**(5), 403–410 (1991)
5. Hou, X., et al.: Dual adaptive pyramid network for cross-stain histopathology image segmentation. In: Shen, D., et al. (eds.) MICCAI 2019. LNCS, vol. 11765, pp. 101–109. Springer, Cham (2019). https://doi.org/10.1007/978-3-030-32245-8_12
6. Isola, P., Zhu, J.Y., Zhou, T., Efros, A.A.: Image-to-image translation with conditional adversarial networks. In: Proceedings of the IEEE Conference on Computer Vision and Pattern Recognition, pp. 1125–1134 (2017)
7. Reinhard, E., Adhikhmin, M., Gooch, B., Shirley, P.: Color transfer between images. IEEE Comput. Graphics Appl. **21**(5), 34–41 (2001)
8. Roux, L., et al.: Mitos & atypia. Image Pervasive Access Lab (IPAL), Agency for Science, Technology and Research Institute for Infocomm Research, Singapore, Technical report 1, pp. 1–8 (2014)
9. Roux, L., et al.: Mitosis detection in breast cancer histological images an ICPR 2012 contest. J. Pathol. Inf. **4**, 8 (2013)
10. Sebai, M., Wang, X., Wang, T.: MaskMitosis: a deep learning framework for fully supervised, weakly supervised, and unsupervised mitosis detection in histopathology images. Med. Biol. Eng. Comput. **58**, 1603–1623 (2020)
11. Vahadane, A., et al.: Structure-preserved color normalization for histological images. In: 2015 IEEE 12th International Symposium on Biomedical Imaging (ISBI), pp. 1012–1015. IEEE (2015)
12. Veta, M., et al.: Predicting breast tumor proliferation from whole-slide images: the tupac16 challenge. Med. Image Anal. **54**, 111–121 (2019)

Domain Generalisation for Mitosis Detection Exploting Preprocessing Homogenizers

Sahar Almahfouz Nasser$^{(\boxtimes)}$ ⓘ, Nikhil Cherian Kurian ⓘ, and Amit Sethi ⓘ

Indian Institute of Technology Bombay, Mumbai, Maharashtra, India
sahar.almahfouz.nasser@gmail.com

Abstract. The detection of mitotic figures in histological tumor images plays a vital role in the decision-making of the appropriate therapy. However, tissue preparation and image acquisition methods degrade the performances of the deep learning-based approaches for mitotic figures detection. MIDOG challenge addresses the domain-shift problem of this detection task. In an endeavour to reduce this domain shift, we propose a pre-processing autoencoder that is trained adversarially to the sources of domain variations. The output of this autoencoder, exhibiting a uniform domain appearance, is finally given as input to the retina-net based mitosis detection module.

Keywords: Domain Generalization · Mitotic Figures · Histopathology · Homogenizers

1 Introduction

Machine learning algorithms often underperform when they are validated on an external data that differs significantly from the distribution of their training data. This problem is even more pronounced in medical images due to several intrinsic sources of variability. The MIDOG challenge presents the problem of domain shift in data for mitosis detection on large cohorts of histopathology dataset collected from several scanners. Mitosis detection by itself is a very challenging problem owing to the large variability in the morphology of mitotic nuclei, along with the presence of several confounding nuclei. This difficulty is further exacerbated by the large scale inter-observer variabilities. In order to reduce the domain discrepancy, we present a preprocessing pipeline that acts as an unsupervised domain generalizer that averages the appearance between the different scanners with an additional capability to nullify domain specific signals. This deep learning pipeline leverages the property of auto-encoders as a cross-data homogenizer, essentially reducing the appearance between the different domains [2].

© Springer Nature Switzerland AG 2022
M. Aubreville et al. (Eds.): MIDOG 2021/MOOD 2021/L2R 2021, LNCS 13166, pp. 77–80, 2022.
https://doi.org/10.1007/978-3-030-97281-3_12

2 Materials and Methods

Our algorithm was trained on MIDOG dataset only. The algorithm consists of a homogenizer followed by RetinaNet [4] for object detection.

2.1 Dataset

The MIDOG challenge released samples obtained from four slide scanners systems, namely the Hamamatsu XR NanoZoomer 2.0, the Hamamatsu S360, the Aperio ScanScope CS2, and the Leica GT450. Around 50 scans were provided from each of these scanner. The entire training data hence consisted of 200 whole slide images WSI from human breast cancer tissue samples stained with routine hematoxylin and eosin (H&E) stain. Supervised training annotations were provided for three scanners except the LeicaGT450. The supervision consisted of mitotic figures and hard negatives that resembles mitotic nuclei. Annotations were collected from multiple experts who were blinded to one another. The preliminary test set contains five WSI correspond to four unrevealed scanners of which only two were also part of the training set. Evaluation of the algorithms was accomplished based on this preliminary test before publishing preliminary results on a leaderboard. The final test consists of 80 cases (20 for each scanner) belonged to the same four scanners used for the preliminary test set.

2.2 Methodology

As mentioned earlier, the object detection was done using RetinaNet, whereas our main contribution lies in proposing and testing a domain generalising preprocessing step. The preprocessing pipeline consists of a multi-headed encoder network G_f. The encoder coupled with a decoder component G_r completes the autoencoder section of the pipeline that aims to reconstruct the input images, $x \in X$ with an MSE loss. The optimization in the MSEloss results in reconstruction of the images that hold an average appearance compared to the training images. Utilizing this idea, we use the whole dataset, with the appropriate validation splits, provided as a part of MIDOG challenge in order to make the autoencoder learn all the latent domains present in data. This is feasible as this part of the training does not require an associated supervised label.

The encoder network also has a training adversarial head G_y which basically acts as a domain discriminator. The reason for incorporating this module is to further erase domain specific signals explicitly with the help of an adversarial component. Here the training process makes use of explicit domain labels in the form of the scanner technology labels, $y \in Y = \{$HamamatsuRx, HamamatsuS360, Aperio, Leica GT450$\}$. We have summarized the overall architecture in Fig. 1.

Mathematically, if we denote the output of the encoder as a D-dimensional feature vector f, For every input x, the outputs of the model are the reconstructed image r and the domain label y.

$$f = G_f(x, \theta_f) \tag{1}$$

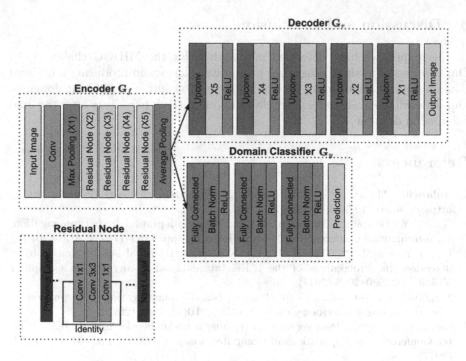

Fig. 1. The architecture of the domain homogenizer.

Finally, during the learning stage, we aim to maximize the domain label prediction loss and minimize the reconstruction loss simultaneously to obtain domain-invariant features.

$$L = \sum_{i=1...N} L_r(G_r(G_f(x_i; \theta_f); \theta_r), y_i) \tag{2}$$

2.3 Network Training

From the data of each of the four scanners we selected 40 WSI out of 50 WSI for training the homogenizer and the remaining 10 WSI used for validation. We used a patch size of 256×256 pixels and a batch size of 8. Furthermore, we performed data augmentation with color jitter, affine transformations, and random lightning and contrast change. We trained the network with a cyclical maximal learning rate [5] of 10^{-4} for 60 epochs until convergence. The loss of the homogenizer was a weighted combination of the perceptual loss and the classification loss. And the model optimized by minimizing the reconstruction loss and maximizing the domain classification loss. For object detection, we followed the same strategy of splitting the data of each of the three annotated scanners into 40 out of 50 WSI of each scanner for training and the remaining 10 WSI for validation. We used the focal loss as the classification loss [3] and L1 loss for regression. We trained the network with a learning rate of 10^{-4} for 150 epochs until convergence.

3   Discussion and Conclusion

In this paper, we have described our method for the MIDOG challenge [1]. Our proposed domain homogenizer proved its efficiency in producing a uniform domain appearance of input images belonging to sources of different domains. Our code will be made publicly available in our GitHub repository after the final submission deadline.

References

1. Aubreville, M., et al.: Mitosis domain generalization challenge. Zenodo (2021). https://doi.org/10.5281/zenodo.4573978
2. Ganin, Y., Lempitsky, V.: Unsupervised domain adaptation by backpropagation. In: International Conference on Machine Learning, pp. 1180–1189. PMLR (2015)
3. Lin, T.Y., Goyal, P., Girshick, R., He, K., Dollár, P.: Focal loss for dense object detection. In: Proceedings of the IEEE International Conference on Computer Vision, pp. 2980–2988 (2017)
4. Marzahl, C., et al.: Deep learning-based quantification of pulmonary hemosiderophages in cytology slides. Sci. Rep. **10**(1), 1–10 (2020)
5. Smith, L.N.: Cyclical learning rates for training neural networks. In: 2017 IEEE Winter Conference on Applications of Computer Vision (WACV), pp. 464–472. IEEE (2017)

Cascade R-CNN for MIDOG Challenge

Salar Razavi[1]([✉]), Fariba Dambandkhameneh[1], Dimitri Androutsos[1],
Susan Done[2,3], and April Khademi[1]

[1] Image Analysis in Medicine Lab (IAMLAB), Electrical, Computer and Biomedical
Engineering, Ryerson University, Toronto, ON, Canada
{salar.razavi,akhademi}@ryerson.ca
[2] Princess Margaret Cancer Centre, University Health Network,
Toronto, ON, Canada
[3] Department of Laboratory Medicine and Pathobiology,
University of Toronto, Toronto, ON, Canada

Abstract. Mitotic counts are one of the key indicators of breast cancer
prognosis. However, accurate mitotic cell counting is still a difficult problem and is labourious. Automated methods have been proposed for this
task, but are usually dependent on the training images and show poor
performance on unseen domains. In this work, we present a multi-stage
mitosis detection method based on a Cascade R-CNN developed to be
sequentially more selective against false positives. On the preliminary
test set, the algorithm scores an F_1 score of 0.7492.

Keywords: Domain Generalization · Mitosis Detection ·
Histopathology · MIDOG

1 Introduction

Breast cancer disease is a global concern affecting over 2 million women worldwide [5]. Central to breast cancer diagnosis and treatment planning, is pathological analysis of tissue sections under magnification. There are typically three features that are used in grading, which includes mitosis counts, tubule formation
and nuclear pleomorphism. Mitotic counts are a key indicator of tumour aggressiveness, but manual counting of mitosis is labourious, subjective and error prone.
With the advent of whole slide imaging (WSI) scanners, there is an opportunity
to leverage computational algorithms to perform mitosis detection in an automatic and objective manner. However, a challenge with automated mitosis detection, and many computational pathology algorithms is domain shift [10]. Different
scanners and staining creates variability in colours and noise distributions, which
can cause generalization challenges especially for deep learning-based algorithms
when new data is out of the training distribution. To address these challenges,
the MIDOG competition was launched [1] to test mitosis detection algorithms on
images acquired from different scanners and laboratories. In addition to the generalization challenges due to domain shift, discriminating mitotic figures from hard
negative samples is also a big concern in mitosis detection. The proposed Cascade
R-CNN based architecture is trained with images from different domains to detect
mitotic figures and hard negative samples with high accuracy.

© Springer Nature Switzerland AG 2022
M. Aubreville et al. (Eds.): MIDOG 2021/MOOD 2021/L2R 2021, LNCS 13166, pp. 81–85, 2022.
https://doi.org/10.1007/978-3-030-97281-3_13

Material and Methods

In this work, the proposed mitosis detection algorithm was developed using the official training set of the MIDOG dataset. The algorithm is based on a publicly available implementation of the Cascade R-CNN [2] which consists of a sequence of sequential detectors with increasing intersection over union (IoU) to reduce false positives which may be attributed to the hard to detect mitotic cells. Because of small amount of images, progressively resampling in each stage is also used to reduce overfitting by ensuring there is a positive set of examples in each stage. These methods and datasets are detailed next.

1.1 Dataset

In total, the MIDOG dataset contains 150 annotated high power fields (HPFs). We extracted 18,960 patches of 512×512 size from three scanners (Hamamatsu XR NanoZoomer 2.0, Hamamatsu S360, Aperio ScanScope CS2). Annotations consisted of two labels: one for the mitotic figures, and a second label for the hard-negative examples, which are darkly stained cells or regions that have similar appearance to mitosis (but are not mitosis). Only patches that had mitosis or hard-negative samples were used to train the models. In total, there were 3,072 training and 1,913 validation patches of 512×512 size, respectively. The data was split randomly. In this subset of data, there were 2,437 mitotic figures and 1,558 hard negative examples. Considering the hard negative examples as an another class ensured the model to reduce the number of false positives specifically in dark-stained and low contrast images.

1.2 Cascade R-CNN

In this work, the Cascade R-CNN architecture [2] is proposed for mitosis detection. The Cascade R-CNN model is a two-stage model that detects candidate regions (region proposal network), and a second stage that performs classification on the candidate regions (RPN+classification). The multi-scale nature of the Cascade R-CNN enables the detection of multiresolution structures by training with increasing IoU thresholds, which may be more robust against false positives. Progressively sampling stage by stage improves detection and ensures that all detectors have a positive set of examples of equivalent size, and as a result reduces overfitting. Applying the same multi-stage procedure in the testing phase, enables a higher agreement between the hypotheses and the detector in each stage.

1.3 Network Training

As previously described, 3,072 patches of size 512×512 pixels with a batch size of 4 were used for training. All images were normalized using the Macenko stain normalization algorithm [7]. Data augmentation with random flipping, scaling, color, cropping and contrast was also considered. Through experimentation, it was found that random cropping (with minimum of 0.3 IoU of cropped patches) and scaling (to 4 other scales with ±64 pixels steps) was optimal (See Fig. 1). The two-stage Cascade R-CNN with ResNext101_64x4d [12] backbone pretrained on ImageNet [4] dataset model with stochastic gradient descent (SGD) and a learning rate of 0.01 for 50 epochs. A linear warm up ratio of 0.001 for 500 steps was also applied to make training more stable. Gradient clipping was considered to prevent exploding gradients. To optimize training, different sampling methods were also considered. The IoU balanced [11] and online hard example mining (OHEM) [9] sampling methods were implemented to select hard samples according to their confidence. However, in comparison with random sampling performance worsened. This may be due to the similarities between mitotic figures (with high score) with some of the hard-negative annotations; or it could be related to the structure of the proposed two-stage detection method. As the Cascade R-CNN model is a two-stage detection (RPN+classification) architecture, added a focal loss [6] to overcome class imbalance but results did not improve and therefore, was not used. positive sample. As the model processes overlapping tiles, there may be multiple detections for a single mitosis. To overcome this, the output predictions were post-processed with non-maximum suppression [8] and a 0.5 threshold is used to remove multiple overlapped bounding box detections. All the models are implemented with MMDetection [3] library for automated detection on a RTX 2080 GTI GPU.

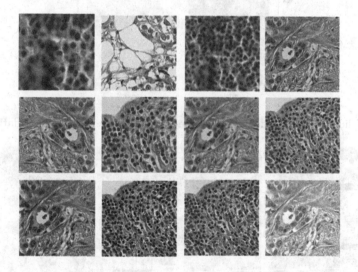

Fig. 1. Sample augmented training images.

2   Evaluation and Results

The precision-recall curves for the validation set are shown in Fig. 2 with an average precision (AP) of 0.8306 for the mitosis class and an AP of 0.6439 for the hard negative class. The average F_1 score on the validation set was 0.63 and 0.46 for the mitosis and hard-negative samples, respectively. Evaluation on the preliminary test set from the MIDOG organizers resulted in a mean F_1 score of 0.7492 (with 0.7707 precision and 0.7289 recall) (See Fig. 3 for some visual results).

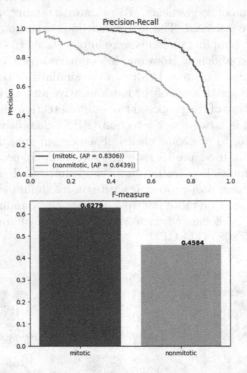

Fig. 2. Validation area under the precision-recall curve per class for validation set.

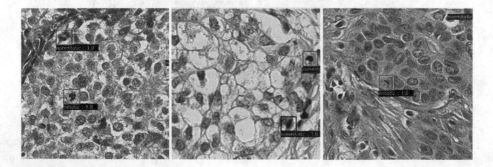

Fig. 3. Results of the detection module at patch-level. Yellow boxes highlights the true predictions, whereas red box shows the false predictions and white boxes are ground-truth annotations.

3   Discussion and Conclusion

In this work, we presented an algorithm for the MIDOG challenge with a F_1 score of 0.6279 on the validation set and an F_1 score of 0.7492 on the preliminary test images. The model's performance on all of the preliminary test images are in the top but only for 003.tiff there are lots of false positives which degraded the overall performance.

References

1. Aubreville, M., et al.: Mitosis domain generalization challenge. https://doi.org/10.5281/zenodo.4573978
2. Cai, Z., Vasconcelos, N.: Cascade R-CNN: delving into high quality object detection (2017)
3. Chen, K., et al.: MMDetection: open MMLab detection toolbox and benchmark. arXiv preprint arXiv:1906.07155 (2019)
4. Deng, J., Dong, W., Socher, R., Li, L.J., Li, K., Fei-Fei, L.: ImageNet: a large-scale hierarchical image database. In: 2009 IEEE Conference on Computer Vision and Pattern Recognition, pp. 248–255 (2009). https://doi.org/10.1109/CVPR.2009.5206848
5. Ferlay, J., et al.: Global cancer observatory: cancer today. Lyon: International agency for research on cancer (2020). https://gco.iarc.fr/today. Accessed Feb 2021
6. Lin, T.Y., Goyal, P., Girshick, R., He, K., Dollár, P.: Focal loss for dense object detection (2018)
7. Macenko, M., et al.: A method for normalizing histology slides for quantitative analysis. In: 2009 IEEE International Symposium on Biomedical Imaging: From Nano to Macro, pp. 1107–1110. IEEE (2009)
8. Neubeck, A., Van Gool, L.: Efficient non-maximum suppression. In: 18th International Conference on Pattern Recognition (ICPR 2006), vol. 3, pp. 850–855. IEEE (2006)
9. Shrivastava, A., Gupta, A., Girshick, R.: Training region-based object detectors with online hard example mining (2016)
10. Stacke, K., Eilertsen, G., Unger, J., Lundström, C.: A closer look at domain shift for deep learning in histopathology. arXiv preprint arXiv:1909.11575 (2019)
11. Wu, S., Yang, J., Wang, X., Li, X.: IoU-balanced loss functions for single-stage object detection (2020)
12. Xie, S., Girshick, R., Dollár, P., Tu, Z., He, K.: Aggregated residual transformations for deep neural networks. arXiv preprint arXiv:1611.05431 (2016)

Sk-Unet Model with Fourier Domain
for Mitosis Detection

Sen Yang[1,2], Feng Luo[3], Jun Zhang[2], and Xiyue Wang[4(✉)]

[1] College of Biomedical Engineering, Sichuan University, Chengdu 610065, China
[2] Tencent AI Lab, Shenzhen 518057, China
[3] Shenzhen International Graduate School, Tsinghua University,
Shenzhen 518131, China
[4] College of Computer Science, Sichuan University, Chengdu 610065, China

Abstract. Mitotic Count is the most important morphological feature of breast cancer grading. Many deep learning-based methods have been proposed but suffer from domain shift. In this work, we construct a Fourier-based segmentation model for mitosis detection to address the problem. Swapping the low-frequency spectrum of source and target images is shown to be effective to alleviate the discrepancy between different scanners. Our Fourier-based segmentation method can achieve F_1 with 0.7456, recall with 0.8072, and precision with 0.6943 on the preliminary test set. Besides, our method reached 1st place in the MICCAI 2021 MIDOG challenge.

Keywords: Mitosis Detection · Deep learning · Domain generalization

1 Introduction

Nowadays, breast cancer is an increasingly common disease in both developed and developing countries [10]. According to the Nottingham grading system [3], it can be diagnosed and predicted by three features, which are nuclear polymorphism, mitotic count, and tubule formation on histopathological sections stained with hematoxylin and eosin (H&E). Among them, the mitotic count is the most important morphological feature of grading. So pathologists usually search for mitosis in a complete slide with a high-power field of view (HPF) manually to count. However, a large number of HPF in a single complete slide and the appearance difference of mitotic cells make the task time-consuming and tedious. In addition, it is objective to judge mitotic cells and are prone to reach a consensus on mitotic count among pathologists.

Recent advances in deep learning and digital scans have paved the way and many automatic mitosis detection methods have been proposed [7,8,11]. Although achieving great success, a drop in performance is often observed when the trained model is tested on data from another domain(i.e. different slide scanners and sample preparation from clinical centers). This problem makes it hard for mitosis detection algorithms to be widely used in the real diagnosis process.

© Springer Nature Switzerland AG 2022
M. Aubreville et al. (Eds.): MIDOG 2021/MOOD 2021/L2R 2021, LNCS 13166, pp. 86–90, 2022.
https://doi.org/10.1007/978-3-030-97281-3_14

To solve the problem, we construct a Fourier-based segmentation method for mitosis detection and submit it to the MIDOG challenge. We convert teh mitosis detection task to mitotic cell segmentation, which makes our model more robust and stable. Inspired by [13], We swap the low-frequency spectrum of source and target images to alleviate the discrepancy between different scanners. Experimental results show that our Fourier-based segmentation method can address the domain shift in mitosis detection. It achieves F_1 with 0.7456, recall with 0.8072, and precision with 0.6943 on the preliminary test set of MIDOG challenge.

2   Methodology

Regarding mitosis detection as segmentation, the proposed algorithm can be divided into image pre-processing, Fourier domain SK-Unet and image post-processing. Image pre-processing and image post-processing are processes of converting bounding boxes and masks. As for the network, SK-Unet [12] equipped with Fourier domain adaptation is modified for the task.

2.1   Image Pre-processing

Due to the fact that the segmentation model is more robust, we convert mitotic detection to segmentation, thus masks of mitotic cells are required. First, all cells in an image are segmented with pre-trained Hovernet [5] which is publicly available[1]. Then we get cells that need to be preserved according to the bounding boxes of the image. In specific, a cell is reserved when the Intersection of Union (IOU) of the cell and any bounding box is over 0.8.

2.2   Fourier Domain Sk-Unet

In order to solve the problem of domain adaptation, a simple method for unsupervised domain adaptation is adopted, which is swapping the low-frequency spectrum of one with the other [13]. To be specific, there are three steps. First, given an image I_s, its amplitude and phase components can be calculated using the FFT algorithm [4]. Second, the center region of I_s's amplitude component is replaced by that of another image I_t. This means that the low-frequency information of the two images is swapped. Third, the modified amplitude component and its unaltered phase component are used to reconstitute an image with a similar style of I_t using inverse FFT (iFFT). The motivation of the swapping process is that high-level semantics represented by the high-frequency spectrum is the real cue for mitosis while low-level semantics is closer to background information. So combining one high-frequency spectrum with several low-frequency components can generate images with different styles and the same label, which enlarges the amount of training data and enhances the generalization ability of

---

[1] https://github.com/simongraham/hovernet_inference.

our model. Some generated samples are shown in Fig. 1. For the mitosis segmentation, SK-Unet is adopted by us. The method proposed a combination of
feature maps from different scales in the encoder-decoder network to improve
the segmentation results.

Fig. 1. A FDA sample. Images are the source image, reference image and generated
image from left to right.

2.3   Post-processing

The image post-processing process aims to refine the result of cell segmentation
and convert it to bounding boxes. Initially, the hole filling technique is applied to
attain accurate segmentation masks. Then, connected component analysis for all
the obtained masks is performed and each connected component is regarded as a
cell. Last, centers of all minimum bounding rectangles for connected components
are calculated as our final result.

3   Experiment

3.1   Dataset

Our algorithm is evaluated on the MICCAI 2021 MIDOG challenge [1]. The
MIDOG training subset consists of 200 Whole Slide Images (WSIs) from
human breast cancer tissue with four slide scanning systems (Hamamatsu XR
NanoZoomer 2.0, the Hamamatsu S360, the Aperio ScanScope CS2, and the
Leica GT450). Each scanner has 50 images annotated, except for the Leica
GT450. To validate the model, we randomly select 50 images from one of the
scanners and model selection is based on models' performance on this validation
set. The rest of images were used to train the model. In addition, there is a
preliminary test set from the MIDOG challenge to evaluate the prior model. It
contains 20 images, which are from 2 scanners in the training set and 2 unknown
scanners.

3.2   Experiment Setup

A sliding window scheme with overlap is used to crop each WSI into small
patches of size 512×512 pixels. Standard real-time data augmentation methods
such as horizontal flipping, vertical flipping, random rescaling, random cropping,
and random rotation are performed to make the model invariant to geometric

perturbations. Moreover, RandomHSV is also adopted to randomly change the hue, saturation, and value of images in the hue-saturation-value (HSV) color space, making the model robust to color perturbations. The Adam optimizer [6] is used as the optimization method for model training. The initial learning rate is set to 0.0003, and reduced by a factor of 10 at the 30th and the 50th epoch, with a total of 80 training epochs. The min-batch size is set as 24. The network is trained by minimizing a total loss function composed of a Focal Loss and a Dice loss. All models are implemented using the PyTorch framework [9] and all experiments are performed on a workstation equipped with an Intel(R) Xeon(R) E5-2680 v4 2.40GHz CPU and four 32 GB memory NVIDIA Tesla V100 GPU cards.

3.3   Experiment Results

The performances of the three models are reported in Table 1. The first two rows are results of LinkNet [2] and SK-Unet [12] respectively. It shows the SK module in SK-Unet indeed gets more informative feature maps in both spatial and channel-wise space than LinkNet. Comparing the second and the third row, our model(SK-Unet+FDA) achieves stronger performance on both the validation set and preliminary test set. It outperforms SK-Unet 0.0141 and 0.0111 of F1-score, which indicates that FDA (Fourier Domain Adaptation) enhances the generalization ability of our model.

Table 1. F1-score from models

Model	F1 (validation set)	F1 (preliminary test set)
LinkNet	0.6954	/
SK-Unet	0.7331	0.7354
Ours	0.7472	0.7465

References

1. Aubreville, M., et al.: Mitosis domain generalization challenge. Zenodo (2021). https://doi.org/10.5281/zenodo.4573978
2. Chaurasia, A., Culurciello, E.: LinkNet: exploiting encoder representations for efficient semantic segmentation. In: 2017 IEEE Visual Communications and Image Processing (VCIP), pp. 1–4. IEEE (2017)
3. Elston, C.W., Ellis, I.O.: Pathological prognostic factors in breast cancer. I. The value of histological grade in breast cancer: experience from a large study with long-term follow-up. Histopathology **19**(5), 403–410 (1991)
4. Frigo, M., Johnson, S.G.: FFTW: an adaptive software architecture for the FFT. In: Proceedings of the 1998 IEEE International Conference on Acoustics, Speech and Signal Processing, ICASSP 1998 (Cat. No. 98CH36181), vol. 3, pp. 1381–1384. IEEE (1998)

5. Graham, S., et al.: Hover-net: simultaneous segmentation and classification of nuclei in multi-tissue histology images. Med. Image Anal. **58**, 101563 (2019)
6. Kingma, D.P., Ba, J.: Adam: a method for stochastic optimization. arXiv preprint arXiv:1412.6980 (2014)
7. Li, C., Wang, X., Liu, W., Latecki, L.J.: DeepMitosis: mitosis detection via deep detection, verification and segmentation networks. Med. Image Anal. **45**, 121–133 (2018)
8. Li, C., Wang, X., Liu, W., Latecki, L.J., Wang, B., Huang, J.: Weakly supervised mitosis detection in breast histopathology images using concentric loss. Med. Image Anal. **53**, 165–178 (2019)
9. Paszke, A., et al.: Automatic differentiation in PyTorch (2017)
10. Saha, M., Chakraborty, C., Arun, I., Ahmed, R., Chatterjee, S.: An advanced deep learning approach for Ki-67 stained hotspot detection and proliferation rate scoring for prognostic evaluation of breast cancer. Sci. Rep. **7**(1), 1–14 (2017)
11. Sebai, M., Wang, X., Wang, T.: MaskMitosis: a deep learning framework for fully supervised, weakly supervised, and unsupervised mitosis detection in histopathology images. Med. Biol. Eng. Comput. **58**, 1603–1623 (2020)
12. Wang, X., et al.: SK-Unet: an improved U-Net model with selective kernel for the segmentation of LGE cardiac MR images. IEEE Sens. J. **21**(10), 11643–11653 (2021)
13. Yang, Y., Soatto, S.: FDA: Fourier domain adaptation for semantic segmentation. In: Proceedings of the IEEE/CVF Conference on Computer Vision and Pattern Recognition. pp. 4085–4095 (2020)

MOOD

MOOD 2021 Preface

Despite overwhelming successes in recent years, progress in the field of biomedical image computing still largely depends on the availability of annotated training examples. This annotation process is often prohibitively expensive because it requires the valuable time of domain experts. Additionally, this approach simply does not scale well: whenever a new imaging modality is created, the acquisition parameters change. Even something as basic as the target demographic is prone to change, and new annotated cases have to be created to allow methods to cope with the resulting images. Image labeling is thus bound to become a major bottleneck in the coming years. Furthermore, it has been shown that many algorithms used in image analysis are vulnerable to Out-of-Distribution samples, resulting in incorrect and overconfident decisions. In addition, physicians can overlook unexpected conditions in medical images, often termed 'inattentional blindness'. In one study 50% of trained radiologists did not notice a gorilla image rendered into a lung CT scan when assessing lung nodules. One approach, which does not require labeled images and can generalize to unseen pathological conditions, is Out-of-Distribution or anomaly detection (which in this context is used interchangeably). Anomaly detection can recognize and outline conditions that have not been encountered during training. Thus, it circumvents the time-consuming labeling process and can quickly be adapted to new modalities. Additionally, by highlighting such abnormal regions, anomaly detection can guide physicians' attention to otherwise overlooked abnormalities in a scan and potentially improve the time required to inspect medical images.

However, while there is a lot of recent research on improving anomaly detection, especially with a focus on the medical field, a common dataset/benchmark to compare different approaches was still missing. Thus, a fair comparison between different proposed approaches was impossible. We tackled this issue for medical imaging with the Medical Out-of-Distribution Analysis Challenge (MOOD 2021) and offered a standardized dataset that allowed for a fair comparison of different approaches and outlined how well different approaches worked.

For the challenge[1], we provided two datasets with more than 600 scans each, one brain MRI-dataset and one abdominal CT-dataset, to allow for a comparison of the generalizability of the approaches. To prevent overfitting on the (types of) anomalies existing in our test set, the test set was kept confidential at all times. The training set consisted of hand-selected scans in which no anomalies were identified. Some scans in the test set did not contain anomalies, while others contained naturally occurring anomalies. In addition to the natural anomalies, we added synthetic anomalies. The target for the challenge was to detect and localize anomalies. As detection and localization are two distinct objectives, the challenge consisted of two (sub-)tasks, first, the detection of abnormal samples, and second, the localization of abnormal pixels.

---

[1] Official challenge document: https://zenodo.org/record/4573948.

The submission was docker-based and ran over the synapse.org[2] platform. To check the submission for validity, the participants had the chance to submit a container for a preliminary evaluation on four previously provided toy cases and validate the reported scores. For participation in the two challenge tasks, three submissions were allowed with only the last one counting towards the challenge. After submission, only the number of successfully processed scans was reported and there was no feedback regarding the score before the official results announcement. From 150 registered participants, for each task, eight teams submitted a valid container. All successfully participating teams were invited to provide a detailed paper describing their approach. The papers were reviewed by the challenge organizers. This resulted in five papers being published in these challenge proceedings.

Thanks go to all participants and the organizing team of MOOD 2021.

David Zimmerer

2 Challenge submission page: https://www.synapse.org/#!Synapse:syn21343101/wiki/599515.

MOOD 2021 Organization

General Chair

David Zimmerer — Medical Image Computing (MIC), German Cancer Research Center (DKFZ), Germany

Program Committee

Tobias Roß — Computer Assisted Medical Interventions (CAMI), German Cancer Research Center (DKFZ) and Quality Match GmbH, Germany

Tim Adler — Computer Assisted Medical Interventions (CAMI), German Cancer Research Center (DKFZ), Germany

Jens Petersen — Medical Image Computing (MIC), German Cancer Research Center (DKFZ), Germany

Gregor Köhler — Medical Image Computing (MIC), German Cancer Research Center (DKFZ), Germany

Paul Jäger — Interactive Machine Learning (IML), German Cancer Research Center (DKFZ), Germany

Peter Full — Heidelberg University, Germany

Klaus Maier-Hein — Medical Image Computing (MIC), German Cancer Research Center (DKFZ), Germany

Steering Committee

Annika Reinke — Computer Assisted Medical Interventions (CAMI), German Cancer Research Center (DKFZ), Germany

Lena Maier-Hein — Computer Assisted Medical Interventions (CAMI), German Cancer Research Center (DKFZ), Germany

Self-supervised 3D Out-of-Distribution Detection via Pseudoanomaly Generation

Jihoon Cho, Inha Kang, and Jinah Park

Computer Graphics and Visualizations Lab., School of Computing,
Korea Advanced Institute of Science and Technology, Daejeon, South Korea
{zinic,rkswlsj13,jinahpark}@kaist.ac.kr

Abstract. Recent studies on anomaly detection have achieved great success in data analysis, yet the application of out-of-distribution detection in medical imaging remains an underdeveloped area of study. In this paper, we propose a 3D fully self-supervised learning method for volumetric medical image data. Inspired by recent advancements in representation learning for out-of-distribution detection, we propose a training method for pseudoanomaly generation with *copy-paste*. The training uses contrasts of the normal image with the pseudoanomaly image that is generated from the normal image. Through this scheme, a representation is learned to detect an abnormal image and to localize the anomaly area. In addition, we use a 3D patch as an input to provide the spatial information of the third dimension from volumetric image data. The proposed approach was tested in the 2021 MICCAI MOOD challenge, and it ranked the first place in both sample-level and pixel-level tasks.

Keywords: Out-of-Distribution · Self-supervised Learning · Medical Imaging

1 Introduction

Creating medical image data with labels for various types of lesions or tumors is a difficult and expensive task since it requires a considerable amount of time by experts in the field. Moreover, even well-trained radiologists often fail to detect abnormalities in unexpected situations. In fact, when radiologists focused on a lung nodule in an image synthesized from a gorilla image into a CT scan, more than half of the radiologists did not detect the gorilla [5]. Even if supervised learning is enabled with a labeled anomaly dataset, it does not work when tested with the data not included in the training dataset [7]. In this regard, we propose a self-supervised learning method that works on unseen datasets when only a normal image is provided.

This research was supported by the Capacity Enhancement Program for Scientific and Cultural Exhibition Services through the National Research Foundation of Korea (NRF) funded by the Ministry of Science and ICT (NRF-2018X1A3A1069693).
J. Cho and I. Kang—Contributed equally to this work as first authors.

M. Aubreville et al. (Eds.): MIDOG 2021/MOOD 2021/L2R 2021, LNCS 13166, pp. 95–103, 2022.
https://doi.org/10.1007/978-3-030-97281-3_15

We train the network with the idea of forging an anomaly, which we called it pseudoanamaly, through hard augmentation. We focus on the fact that most of the features of the original image are lost during the augmentation process due to strong distortion [3]. Thus, we observe that if strong distortion, called hard augmentation, is applied to a partial patch of the image, the patch itself becomes an out-of-distribution area. Among the various types of hard augmentations, *copy-paste*, which is a method of pasting copied patches to different areas, shows robust and effective performances for various cases. Therefore, we set our strategy to develop self-supervised learning by pseudoanomaly generation with *copy-paste* by combining additional subaugmentations to resolve unexpected dependencies.

In addition, most of the existing methods are trained with 2D slices [2,17], but these approaches lose much of the context information of 3D CT scans. Moreover, the computation is too expensive for a 3D image to obtain a precise result since a pixel-level abnormality is generated based on a sample-level abnormality [9,19]. Therefore, we devise a 3D encoder-decoder network in which the attached classification module computes an overall abnormality and the decoder generates a probability map of abnormalities. When constructing the network model, the characteristics of the target domain are considered. In the case of complex structured data, considering the fact that there are different anatomical features according to the different portions, we design the network to learn each positional characteristic through the position module. In contrast, in the case of regularly structured data, considering that there is little difference in appearance depending on the position, we process the entire image at once without the position information. Then, we solve the inconsistent abnormality problem using the consistency connection module, which allows pixel-level results to reflect the sample-level result.

The experimental results of the proposed method were presented in the 2021 MICCAI MOOD challenge [20], and this method ranked in the first place for both the pixel-level task and the sample-level task.

2 Related Work

Self-supervised learning has drawn increasing attention in recent years when supervised learning encountered the bottleneck of expensive data labeling. Autoencoder models [1], which are used as generative self-supervised learning methods, train the data distribution by reconstructing the image from the corrupted input. Although it generates the classification result from a pointwise generative reconstruction, it also has the shortcoming that the result is highly dependent on the intensity difference. Another self-supervised learning method is contrastive learning. In the case of instance discrimination, positive and negative samples need to be defined to train the instance representation itself. Many research works [3,4,6] have selected a scheme that makes the augmented inputs similar and the other image in the batch dissimilar. However, there is an issue that some augmentations can degrade the discriminative performance [3], so CSI [16] has been proposed to treat hard augmentations as negative samples.

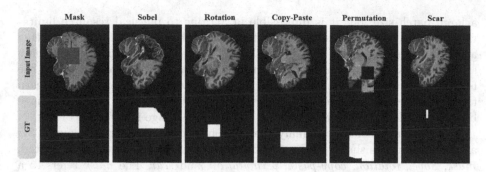

Fig. 1. Examples of Hard Augmentation. The size and position of the patch to be augmented are randomly selected.

Anomaly detection is done by self-supervised learning. The task is to identify abnormal images and to find its location when a dataset consisting only of normal images is provided. The definition of an anomaly image includes those cases ranging from a tiny defect to the distribution of out-of-normal images. Initially, autoencoder-based networks were developed [10,12,14]. These approaches had limitations in increasing the accuracy due to the differences occurring during reconstruction. Another method, two-stage training has been studied; it learns the overall features of normal data through pretasks and then performs anomaly detection through fine-tuning. Path SVDD [19] conducted a pretask to find the relative position of the peripheral patch relative to the given patch in the first stage. CutPaste [9] is an example of creating an irregularity by pasting a copied image patch at the first stage, resulting in generalizing effectively to real defects at the second stage.

MOOD challenge [20] addresses another challenge from the previously studied anomaly detection field in that it conducts the task with 3D medical images. In the previous study of MOOD 2020 [11,18], they proposed self-supervised approaches, which are still 2D image-based approaches that divide the volume image into 2D slices. If only 2D images are considered, they inevitably lose much of the 3D information.

3   Method

We use U-Net [13] as a reference network that receives a 3D patch as an input. The pixelwise abnormality is performed through the decoder, and the samplewise abnormality is performed by attaching a classification module to the bottom of the network.

3.1   Pseudoanomaly Generation

To detect anomalies, training of the normal distribution is required. Existing 2D methods are able to learn the representation of normal instances indirectly with a large batch size, but it is impossible in 3D due to physical resource limitations. Therefore, we choose a straightforward method of arbitrarily generating a pseudoanomaly even if only normal data are given. The processed anomaly area can be said to be an out-of-normal distribution area because most of the features of the original image are lost. Therefore, we train the encoder to learn the difference between the normal and abnormal images from the same image by the classification task to determine whether the image is defective or not.

To create a reasonable anomaly image, we have examined various types of 3D augmentations. As shown in Fig. 1, six types of hard augmentations are tested: *mask*, *Sobel*, *rotation*, *copy-paste*, *permutation*, and *scar*. The *mask* refers to a method of giving a random intensity value $\in [0, 1]$, the *Sobel* refers to a method of applying a Sobel filter to a random patch, and *rotation* refers to a method of rotating a random patch in $[90, 180, 270]$ degrees. *copy-paste* is a method of pasting a copied patch to another area, *permutation* is a method of mixing a random patch with 8 subpatches divided in the x-, y-, and z-axis directions and *scar* [9] is a method of applying *copy-paste* to a very small area.

When the encoder learns the normal distribution, the decoder predicts where the anomaly is located. It is a kind of segmentation task, so it is vulnerable to overfitting to the trained pseudoanomalies. Simple 3D pseudoanomalies lack diversity because they are mostly cuboid in shape and have a limited intensity range. To avoid overfitting problems, we perform additional augmentations of color jitter and free rotation in three dimensions (Fig. 2).

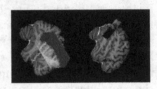

Fig. 2. Examples of Subaugmentation

3.2 Network Module

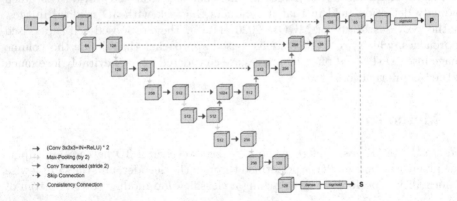

Fig. 3. Network Architecture with Consistency Connection. The number on each cube indicates the size of the channel. **S** indicates the sample-level results, and **P** indicates the pixel-level results. Note the 'consistency connection' indicated in the network.

Consistency Connection. Sample-level and pixel-level anomaly detections have a common goal of finding anomalous images, although these two tasks are performed in different modules. Therefore, a consistency connection is added to increase the learning efficiency by sharing the results of the two tasks during training. As shown in Fig. 3, the classification results are concatenated with the last part of the decoder to the pixel level result to reflect the sample level result. The consistency connection aims to raise the consistency between two tasks that are trained in different modules so that a synergistic effect can be obtained in which the accuracy of both tasks increases.

Position Module. Often the features are differentiated in accordance with their positions. In such case, providing position information during training can be helpful. In particular, human body image includes complex structures that can be interpreted easily by knowing its spatial position. To take advantage of this property, we attach the position information to the input patch by defining the position module, as shown in Fig. 4. Position information contains the defined position class, and it is added to the corresponding channel of the input image patch before training. In this way, the representative features of each position are learned to inform the global context of the network.

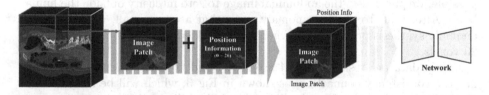

Fig. 4. Scheme of Position Module. This is an example of defining 27 position classes by dividing the input image into three each in the x, y, and z directions. Position information with the corresponding channel to each class is concatenated to the input image patch.

4 Experiments

We evaluate our method using the MOOD 2021 dataset, which consists of 800 brain CT images of $256 \times 256 \times 256$ and 550 abdominal CT images of $512 \times 512 \times 512$. It is divided into a training set, validation set and test set at a ratio of 8:1:1. During the training, elastic deformation was applied to the input image with 50% probability. All models were trained using the Adam [8] optimizer with a one-cycle learning rate policy [15] ranging from 10^{-3} to 10^{-2} for 200 epochs. All of the training was done on a NVIDIA RTX 3090 GPU.

A pseudoanomaly is created by randomly selecting a center position and applying hard augmentation to a region with a random width, height, and depth. Their length ranges from 10 pixels to half of the input size for each dimension. In the case of *scar*, we apply different synthesis rules that select the length of

Table 1. The Test Results of Six Types Hard Augmentation. Models are trained with a synthesized pseudoanomaly by six different hard augmentations without subaugmentation. Numbers written without brackets indicate a pixel-level AP, and numbers in bracket indicate a sample-level AP. Sample-level results of the total AP are described in the order of the results for normal images and for anomaly images. Bold text indicates the method showing the highest AP, and the underlined text indicates the method showing the second-highest AP.

Data	Train	Test						Total AP
		Mask	Sobel	Rotation	Copy-Paste	Permutation	Scar	
Brain	Mask	0.9995 (1.0)	0.2062 (0.0375)	0.1779 (0.2375)	0.2566 (0.2875)	0.1555 (0.0875)	0.1700 (0.0)	0.3276 (1.0, 0.2750)
	Sobel	0.3666 (0.55)	0.0.9801 (1.0)	0.3361 (0.4)	0.3796 (0.55)	0.5043 (0.675)	0.3102 (0.0375)	0.4795 (1.0, 0.5354)
	Rotation	0.9893 (1.0)	0.8901 (0.9625)	0.9454 (0.975)	0.9521 (1.0)	0.8398 (0.9875)	0.5476 (0.4575)	<u>0.8596</u> (0.975, 0.9021)
	Copy-Paste	0.9974 (1.0)	0.9128 (0.9125)	0.9426 (0.95)	0.9900 (1.0)	0.8642 (0.9875)	0.5824 (0.475)	**0.8815** (1.0, 0.8875)
	Permutation	0.8152 (1.0)	0.8382 (0.975)	0.9342 (0.9875)	0.9643 (1.0)	0.9601 (1.0)	0.6198 (0.4875)	0.8553 (1.0, 0.9083)
	Scar	0.4935 (0.9875)	0.5003 (0.8125)	0.6740 (0.975)	0.6244 (1.0)	0.7213 (0.9875)	0.8881 (0.9)	0.6503 (0.975, 0.9438)
Abdom	Mask	0.9518 (1.0)	0.6262 (0.8727)	0.1424 (0.2727)	0.1989 (0.3818)	0.1431 (0.2910)	0.0054 (0.0)	0.3446 (1.0, 0.4697)
	Sobel	0.1770 (0.7818)	0.9811 (1.0)	0.2324 (0.7091)	0.2826 (0.8)	0.3037 (0.7455)	0.1627 (0.2182)	0.3566 (0.8727, 0.7091)
	Rotation	0.5583 (1.0)	0.6581 (1.0)	0.6627 (1.0)	0.6152 (1.0)	0.7604 (1.0)	0.3206 (0.5455)	<u>0.5959</u> (0.8545, 0.9242)
	Copy-Paste	0.9125 (1.0)	0.9290 (1.0)	0.8364 (1.0)	0.9317 (1.0)	0.8954 (1.0)	0.2662 (0.4727)	**0.7952** (0.9091, 0.9121)
	Permutation	0.3832 (1.0)	0.6201 (0.9818)	0.5011 (1.0)	0.6406 (1.0)	0.8416 (1.0)	0.3100 (0.5455)	0.5494 (0.8000, 0.9212)
	Scar	0.1190 (1.0)	0.2452 (1.0)	0.2199 (0.9818)	0.2289 (1.0)	0.3510 (0.9818)	0.6614 (0.8)	0.3042 (0.4909, 0.9606)

the long side from 5 pixels to half of the input size, and the other sides from 2 pixels to 5 pixels. If a pseudoanomaly is generated in the background with the same intensity of air, we do not consider this area as a label. To make it possible, we preprocess the abdominal image to zero intensity outside the human body. After that, to prevent anomaly areas that are too small, we perform the anomaly synthesis process again when the generated pseudoanomaly is less than 20 voxels.

Brain data are resized into $64 \times 64 \times 64$ and trained with the model that has the consistency connection, as shown in Fig. 3, which will be referred to as **BrainNet**. In the case of resizing into $128 \times 128 \times 128$, the detection performance was rather poor. This is because it is difficult to create a larger receptive field due to a limitation in the GPU's VRAM. In the case of abdominal data, it is resized into $256 \times 256 \times 256$ and trained with the model that has the position module of Fig. 4 instead of the consistency connection, which will be called **AbdomNet**. When the position module is used, the network is trained with a patch-based approach in which the input has half of the data size. Both networks take free-rotated and color-jittered *copy-paste* anomalies as inputs. We set the different batch sizes to 8 and 1 for BrainNet and AbdomNet, respectively, considering the input image size. In the experiments, we resized the abdominal image to $128 \times 128 \times 128$ due to training time and resized the brain image to $128 \times 128 \times 128$ when attaching the positional module to BrainNet.[1]

4.1   Augmentation Test

We first compare the effectiveness of six augmentations using a robustness test on the detection of unseen augmentation types. As you can see in Table 1, the network trained with *copy-paste* showed outstanding performance on both the brain

---

[1] Code available at https://github.com/zinic95/MOOD_CGV.

Table 2. The Results of the Ablation Study. BrainNet is a model with a consistency connection and AbdomNet has a position module instead of a consistency connection. Both networks make a pseudoanomaly by free-rotated and color-jittered *copy-paste*. It shows the difference in AP results by excluding the methods of BrainNet and AbdomNet. *pos* indicates the position module and *cons* indicates the consistency connection.

Network	Shape		Size			Intensity				AP
	Cube	Sphere	S	M	L	0.25	0.5	0.75	1.0	
BrainNet	0.9347	0.8782	0.7416	0.9834	0.9962	0.7871	0.8828	0.9706	0.9878	0.9064
	(0.9054)	(0.7751)	(0.6270)	(0.9202)	(0.9289)	(0.6217)	(0.8137)	(0.9059)	(0.9629)	(0.8403)
– *free rotation*	-0.0412	-0.3080	-0.3352	-0.0945	-0.0960	-0.1323	-0.2112	-0.1905	-0.1669	-0.1746
	(-0.0461)	(-0.2780)	(-0.3133)	(-0.0961)	(-0.0327)	(-0.0378)	(-0.2032)	(-0.1880)	(-0.1624)	(-0.1621)
– *color jitter*	-0.0917	-0.1319	-0.2059	-0.0830	-0.0482	-0.0420	-0.0894	+0.0028	-0.0903	-0.1118
	(-0.0722)	(-0.0984)	(-0.0971)	(-0.0722)	(-0.0420)	(+0.0028)	(-0.0735)	(-0.1102)	(-0.1238)	(-0.0853)
– *cons*	-0.0256	-0.0652	-0.0885	-0.0050	-0.0047	-0.0184	-0.0578	-0.0335	-0.0214	-0.0454
	(-0.0558)	(-0.0486)	(-0.0516)	(-0.0537)	(-0.0267)	(-0.0130)	(-0.0418)	(-0.0603)	(-0.0428)	(-0.0522)
– *cons*, + *pos*	-0.1299	-0.2063	-0.0166	-0.0445	-0.4450	-0.1705	-0.1791	-0.1582	-0.1669	-0.1681
	(-0.0027)	(-0.0121)	(+0.0878)	(+0.0013)	(-0.0666)	(+0.0945)	(+0.0107)	(-0.0214)	(-0.0564)	(-0.0074)
AbdomNet	0.7965	0.7161	0.4147	0.9749	0.8796	0.6337	0.7784	0.8017	0.8115	0.7563
	(0.8984)	(0.8577)	(0.7679)	(0.9673)	(0.8989)	(0.7978)	(0.8865)	(0.9070)	(0.9207)	(0.8780)
– *free rotation*	+0.0340	-0.3772	-0.0821	-0.1754	-0.2576	-0.1948	-0.1690	-0.1645	-0.1580	-0.1716
	(-0.0293)	(-0.4077)	(-0.3814)	(-0.1649)	(-0.1092)	(-0.3346)	(-0.2086)	(-0.1756)	(-0.1550)	(-0.2185)
– *color jitter*	-0.0891	-0.1666	-0.0077	-0.1018	-0.2743	-0.0534	-0.1675	-0.1639	-0.1266	-0.1279
	(-0.0236)	(-0.2743)	(-0.1249)	(-0.1361)	(-0.1859)	(-0.1606)	(-0.1647)	(-0.1512)	(-0.1081)	(-0.1490)
– *pos*	0.0000	-0.0011	+0.0554	-0.0048	-0.0526	-0.0317	+0.0194	+0.0026	+0.0075	-0.0006
	(-0.0385)	(-0.2429)	(-0.1984)	(-0.0652)	(-0.1585)	(-0.1983)	(-0.1368)	(-0.1428)	(-0.0829)	(-0.1407)
+ *cons*	+0.0676	+0.0687	+0.1524	+0.0009	+0.0431	+0.0728	+0.0726	+0.0706	+0.0686	+0.0681
	(-0.0110)	(-0.0413)	(-0.1200)	(-0.0060)	(+0.0123)	(-0.0721)	(-0.0309)	(-0.0281)	(-0.0205)	(-0.0261)

and abdominal datasets. In particular, *copy-paste* on the abdominal dataset shows that pixel-level AP is approximately 0.2 higher than the second-place *rotation* case. Since there are no significant differences in the sample-level AP when comparing augmentation cases belonging to the upper rank, we set a strategy to perform the self-supervised learning by making a pseudoanomaly with *copy-paste* according to the pixel-level AP results. Then, we enrich the anomaly shape and intensity range using additional augmentation to the *copy-paste* anomaly area. We apply color jitter and *rotation*, which showed notable performance on the augmentation test, with more free dimensions and degrees.

4.2 Ablation Study

We perform an ablation study on the proposed subaugmentations and network modules. We report the pixel-level AP and sample-level AP of our 3230 synthesized mask anomaly images with various shapes, sizes, and intensities from five test-set images. The results are shown in Table 2.

Subaugmentation. Free rotation significantly improves the performance of sphere-shaped anomalies. This is because *copy-paste* augmentation is mostly shaped as a cuboid and aligned with each image dimension. Since the model learns this shape characteristic of the anomaly area during training, overfitting to the cube shape may occur. Similarly, the color jitter shows better results at almost all intensities. *copy-paste* cannot generate intensity outside the original image intensity range. Color jitter, however, enables the generation of the intensity that the original image does not have, thus avoiding the model being trained only to specific intensities.

Network Module. In the case of BrainNet, the consistency connection improves the performance by directly transferring the classification result to the pixel-level detection. However, the position module greatly degrades the pixel-level performance. The brain has a symmetrical structure, and there is no drastic change because of location so that the positional module harms the learning of the brain structure. In contrast to BrainNet, the position module of AbdomNet greatly improves the sample-level task. The position information helps to understand the complex structure of a human abdomen, which has a better learning effect. When attaching the consistency module to AbdomNet, there is a trade-off between the advantage of pixel-level tasks and the disadvantage of sample-level tasks. Therefore, in light of the results of the provided 4 toy set cases, which have sphere-shaped anomalies with diverse intensities, as shown in Fig. 5, we design the final AbdomNet without a consistency module.

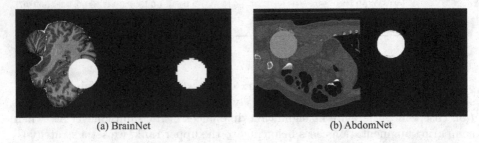

(a) BrainNet   (b) AbdomNet

Fig. 5. Results of the Toy Set. (Left) Input image and (Right) Prediction result.

5   Conclusion

We present a novel self-supervised learning method for 3D volumetric data. The key idea is the creation of a pseudoanomaly to identify the defectivities and their areas. When training on out-of-distribution by *copy-paste* pseudoanomaly generation, we prove that it shows robust results even in unseen cases. Moreover, we designed helpful network modules, which are the consistency connection module and position module, in regards to the characteristics of the data. Our leaning method shows superior results on the MOOD 2021 dataset with BrainNet and AbdomNet for anomaly detection in brain CT images and abdominal CT images, respectively. Our next plan is to build a network to better manage large-sized images without resizing and make the network more robust to diverse anomalies such as image corruptions.

References

1. Ballard, D.H.: Modular learning in neural networks. In: AAAI, vol. 647, pp. 279–284 (1987)
2. Chen, L., et al.: Self-supervised learning for medical image analysis using image context restoration. Med. Image Anal. **58**, 101539 (2019)
3. Chen, T., Kornblith, S., Norouzi, M., Hinton, G.: A simple framework for contrastive learning of visual representations. arXiv preprint arXiv:2002.05709 (2020)
4. Chen, T., Kornblith, S., Swersky, K., Norouzi, M., Hinton, G.: Big self-supervised models are strong semi-supervised learners (2020)
5. Drew, T., Võ, M.L.H., Wolfe, J.M.: The invisible gorilla strikes again: sustained inattentional blindness in expert observers. Psychol. Sci. **24**(9), 1848–1853 (2013)
6. He, K., Fan, H., Wu, Y., Xie, S., Girshick, R.: Momentum contrast for unsupervised visual representation learning (2020)
7. Hendrycks, D., Mazeika, M., Kadavath, S., Song, D.: Using self-supervised learning can improve model robustness and uncertainty (2019)
8. Kingma, D.P., Ba, J.: Adam: a method for stochastic optimization (2017)
9. Li, C., Sohn, K., Yoon, J., Pfister, T.: CutPaste: self-supervised learning for anomaly detection and localization. CoRR abs/2104.04015 (2021). https://arxiv.org/abs/2104.04015
10. Lu, Y., Xu, P.: Anomaly detection for skin disease images using variational autoencoder (2018)
11. Marimont, S.N., Tarroni, G.: Anomaly detection through latent space restoration using vector-quantized variational autoencoders (2020)
12. Perera, P., Nallapati, R., Xiang, B.: OCGAN: one-class novelty detection using GANs with constrained latent representations (2019)
13. Ronneberger, O., Fischer, P., Brox, T.: U-Net: convolutional networks for biomedical image segmentation. CoRR abs/1505.04597 (2015). http://arxiv.org/abs/1505.04597
14. Sabokrou, M., Khalooei, M., Fathy, M., Adeli, E.: Adversarially learned one-class classifier for novelty detection (2018)
15. Smith, L.N., Topin, N.: Super-convergence: very fast training of neural networks using large learning rates. In: Artificial Intelligence and Machine Learning for Multi-Domain Operations Applications, vol. 11006, p. 1100612. International Society for Optics and Photonics (2019)
16. Tack, J., Mo, S., Jeong, J., Shin, J.: CSI: novelty detection via contrastive learning on distributionally shifted instances (2020)
17. Taleb, A., Lippert, C., Klein, T., Nabi, M.: Multimodal self-supervised learning for medical image analysis. In: Feragen, A., Sommer, S., Schnabel, J., Nielsen, M. (eds.) IPMI 2021. LNCS, vol. 12729, pp. 661–673. Springer, Cham (2021). https://doi.org/10.1007/978-3-030-78191-0_51
18. Tan, J., Hou, B., Batten, J., Qiu, H., Kainz, B.: Detecting outliers with foreign patch interpolation. CoRR abs/2011.04197 (2020). https://arxiv.org/abs/2011.04197
19. Yi, J., Yoon, S.: Patch SVDD: patch-level SVDD for anomaly detection and segmentation (2020)
20. Zimmerer, D., et al.: Medical out-of-distribution analysis challenge 2021, March 2021. https://doi.org/10.5281/zenodo.4573948

Self-supervised Medical Out-of-Distribution Using U-Net Vision Transformers

Seongjin Park[✉][iD], Adam Balint[iD], and Hyejin Hwang[iD]

Noul Co., Ltd., Yongin-si, Republic of Korea
{jimmy,adam,lily}@noul.kr

Abstract. Mitigating out-of-distribution inaccuracies due to insufficient labeled data is a widespread problem in the medical AI field. To tackle this medical out-of-distribution problem, through the medical out-of-distribution challenge [1], our team utilized self-supervised learning with UNETR [2]. UNETR is a 3D UNET model where the encoder incorporates Vision Transformers [3]. Abnormal samples were generated from normal samples using a 3D extension to the cutout method described by [4], allowing for self-supervised learning. The input size was chosen to be one eighth of the original size due to the large computational cost associated with 3D image segmentation. This results in 3D images that are $128 \times 128 \times 128$ pixels for brain samples and $256 \times 256 \times 256$ pixels for abdominal samples. Abdominal samples were further divided into 8 different patches, with all 8 patches being used for training. The model produces a voxel-wise abnormality prediction ranging between 0 and 1, where 1 represents an abnormal voxel. We then further extrapolate this prediction to generate a sample-wise prediction by taking the maximum value.

Keywords: Semi-supervised learning · 3D UNETR · Medical Out-of-distribution

1 Introduction

Abnormality causes a great threat to the diagnostic system [1,5–7]. However, acquiring quality and important data from biomedical image is limited especially with abnormal samples. For example, a low frequency of appearance of rare disease or corruption of the data makes data acquisition even more difficult. Also, accurate labeling by professional personnel is costly in terms of time and resource [1]. Thus, detecting such anomaly is a widespread problem in the medical AI field.

Various approaches have been attempted to define and interpret the out-of-distribution problem. First, when a feature is extracted from the classifier, a study was conducted to create a threshold with SVM, binary classifier, probability threshold [8] to distinguish the in-distribution samples from the out-of-distribution samples. Alternatively, there is an approach that utilizes the concepts of temperature scaling and input preprocessing. It can distinguish in-distribution samples and out-of-distribution samples data based on a threshold [9,10]. Also, classification based on distance [11] has been achieved.

© Springer Nature Switzerland AG 2022
M. Aubreville et al. (Eds.): MIDOG 2021/MOOD 2021/L2R 2021, LNCS 13166, pp. 104–110, 2022.
https://doi.org/10.1007/978-3-030-97281-3_16

Semi-supervised anomaly detection has been achieved by setting a discriminative boundary of a normal sample [12,13]. Unsupervised anomaly detection is a method of encoding an unlabeled sample with an auto-encoder, extracting the principal component, calculating difference between the input and the output to determine whether an abnormal sample [14–16]. There is also self-supervised learning methods to solve out-of-distribution problems. By well-defined pre-task definitions, we can use the advantages of supervised learning to overcome the limitations of solving problems with only one normal sample class [17].

To solve the medical out-of-distribution problem, we chose semi-supervised learning method based on the winning paper of the last year [17]. The idea of semi-supervised learning, which generated abnormal samples using normal samples, was extended with the cutout [4]. Also, instead of using 2D image segmentation [18–20], we decided to use 3D UNETR [2]. Among various 3D UNET variants [21,22], 3D UNETR has an advantage at learning global context. In MOOD 2021, we achieved 3rd place and 5th place in pixel-wise task and sample-wise task respectively. However, as we analyzed the results, trained model for abdominal could be improved. We increased the size of the cutout and found out that the result significantly improves.

2   Methods

2.1   Self-supervised Dataset Preparation

The dataset that was provided in the challenge consists of 800 brain MRI scans with the size of $256 \times 256 \times 256$ and 550 abdominal CT scans with the size of $512 \times 512 \times 512$. Additionally, 4 brain and 4 abdominal samples were given as a toy examples. Each of these sets consisted of both abnormal and normal samples. All of the training samples were normal. To acquire abnormal samples, a method similar to self-supervised learning [17] was used, especially a 3D extension to the cutout method [4]. In Fig. 1, normal abdominal and brain data samples are shown in (a) and (d), respectively. For Fig. 1 (b) and (e) show abnormal samples generated using the cutout method, and (c) and (f) show the corresponding label used in the training. Total number of cutout was 5 for each sample.

2.2   3D UNETR

3D UNETR [2] was chosen and utilized for the self-supervised learning task. UNETR follows the "U-shaped" network design. A set of transformer replaces the encoder with convolutions, and the outputs from the transformer are connected to the decoder with skip connections. A transformer turns the input volume into sequence representations, which allows the network to learn the global context effectively. The model structure for 3D UNETR was kept the same from the original paper [2]. The patch resolution was set to be 16, and the embedding size was set to be 768.

Since training 3D models with 3D volume requires more memory than training 2D models with images from a stack, it was necessary to reduce the size of

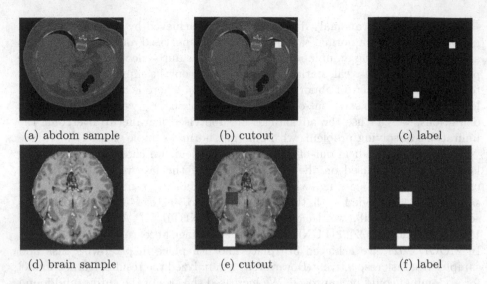

(a) abdom sample   (b) cutout   (c) label

(d) brain sample   (e) cutout   (f) label

Fig. 1. Example of original samples, abnormal samples and their labels for abdominal and brain data

the input. We reduced the size of brain samples to $128 \times 128 \times 128$. For abdominal samples, their size were reduced down to $256 \times 256 \times 256$ instead. Reducing the abnormal samples down to $128 \times 128 \times 128$ caused too much loss in the information. To allow using the abnormal samples as inputs in training, the samples were divided into 8 pieces as shown in Fig. 2. Each individual patch was fed into the network, and the final output was produced by combining all of the 8 output pieces back to the original position. For brain samples, no such patch division was needed.

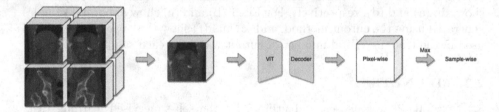

Fig. 2. Model prediction pipeline

Figure 2 shows the model prediction pipeline for both pixel-wise and sample-wise prediction. The first output of the model represents the pixel-wise prediction. Then, the sample-wise prediction is acquired by calculating the maximum value of the pixel-wise prediction, which is from 0 to 1. The training is achieved minimizing the binary cross entropy loss between the pixel-wise prediction and the actual label. To train the model, ADAM optimizer was used with a learning rate of 0.001 and a weight decay of 0.0005. At least 5000 iterations were allowed for training both brain and abdominal samples. 6 GPUs (NVIDIA Quadro RTX 5000) were used.

3 Results and Discussion

Using the toy testing examples that were provided, results listed in Table 1 were achieved. The model provided more accurate result on pixel-wise results except for the abdominal samples. The pixel-wise abdominal results were significantly lower than expected. From Fig. 4 (b), it can be seen that the model was confused on the edges of the divided patches of abdominal. Also, some parts of the bone were misclassified as anomaly. We speculated the size of the cutout compared to the large size of the abdominal samples. Since the same size of the cutout augmentation was applied on the abdominal samples as that of the brain samples, small sized cutout may have prevented the network from learning enough features for anomaly detection, causing low accuracy on pixel-wise prediction for abdominal samples.

Table 1. UNETR accuracies on MOOD2021 sample testing data.

	Sample-wise prediction	Pixel-wise prediction
Brain	0.75	0.78
Abdominal	0.75	0.16
Abdominal improved after challenge	0.75	0.84

To verify our hypothesis, the size of the cutout in the abdominal samples was doubled as shown in Fig. 3. Our preliminary experimental results show that doubling the size of the cutout significantly increases the accuracy from 0.16% to 0.84% on toy examples. Figure 4 demonstrates examples of the visualized prediction results of abnormal samples for a brain model, an abdominal model, and an improved abdominal model. As it can be seen from (b) and (c), the improved model not only has learned better for anomaly but also better identifies the edges of the blocks than the original abdominal model does.

In the MOOD2021 challenge, the variation of out-of-distribution samples was not revealed ahead of time. While, our method only used cutout to generate corrupt examples, some other simple augmentations that could have been implemented, and were later revealed to be part of the hidden testing set, include local shuffling and varying the transition rate between normal and anomalous portions of the image by blending them together.

Based on the research conducted during this competition, we believe that the volumetric segmentation approach is promising for the medical out-of-distribution task. We believe that future exploration of the frequency, intensity and variety of image corruptions will lead to better, more robust out-of-distribution results. To start, implementing local shuffling and blending between

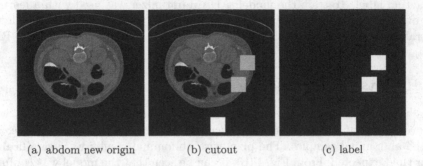

(a) abdom new origin   (b) cutout   (c) label

Fig. 3. Increased size of cutouts applied on abdominal dataset generation for improved abdominal model

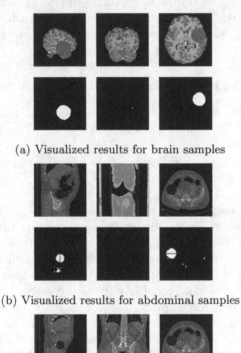

(a) Visualized results for brain samples

(b) Visualized results for abdominal samples

(c) Visualized results for abdominal samples

Fig. 4. Visualization of results

normal and anomalous portions of the image is a good start, however exploring avenues such as the addition of local noise, mixing and matching portions of different brain scans could also be explored. A more in depth approach could be creating a 3D mesh model given the different data scans and applying shape distortions to more accurately model different medical conditions such as swelling in the brain could also lead to fruitful results.

4   Conclusion

To tackle medical out-of-distribution problem, we utilized self-supervised learning with 3D UNETR. We demonstrated that 3D UNETR is effective at learning global context to predict pixel-wise anomaly. Initially from the result, we identified that too small cutout size may hinder the training, and increasing the cutout size significantly increased the model performance on toy examples. Although, we only used the cutout for anomaly generation, other methods of augmentation such as local shuffling or foreign patch interpolation [17] may be applied to achieve better results.

References

1. Zimmerer, D., Petersen, J., Köhler, G., et al.: Medical out-of-distribution analysis challenge 2021, March 2021. https://doi.org/10.5281/zenodo.4573948
2. Hatamizadeh, A., Yang, D., Roth, H., Xu, D.: UNETR: transformers for 3D medical image segmentation. arXiv: 2103.10504 [eess.IV] (2021)
3. Dosovitskiy, A., Beyer, L., Kolesnikov, A., et al.: An image is worth 16 × 16 words: transformers for image recognition at scale. arXiv: 2010.11929 [cs.CV] (2020)
4. DeVries, T., Taylor, G.W.: Improved regularization of convolutional neural networks with cutout a.rXiv preprint arXiv:1708.04552 (2017)
5. Shvetsova, N., Bakker, B., Fedulova, I., Schulz, H., Dylov, D.V.: Anomaly detection in medical imaging with deep perceptual autoencoders. IEEE Access 9, 118 571–118 583 (2021). ISSN: 2169-3536, https://doi.org/10.1109/ACCESS.2021.3107163
6. Nguyen, A., Yosinski, J., Clune, J.: Deep neural networks are easily fooled: high confidence predictions for unrecognizable images. arXiv: 1412.1897 [cs.CV] (2015)
7. Cao, T., Huang, C.-W., Hui, D.Y.-T., Cohen, J.P.: A benchmark of medical out of distribution detection. arXiv: 2007.04250 [cs.LG] (2020)
8. Hendrycks, D., Gimpel, K.: A baseline for detecting misclassified and out-of-distribution examples in neural networks. arXiv: 1610.02136 [cs.NE] (2018)
9. Liang, S., Li, Y., Srikant, R.: Enhancing the reliability of out-of distribution image detection in neural networks. arXiv: 1706.02690 [cs.LG] (2020)
10. Lee, K., Lee, K., Lee, H., Shin, J.: A simple unified framework for detecting out-of-distribution samples and adversarial attacks. arXiv: 1807.03888 [stat.ML] (2018)
11. Napoletano, P., Piccoli, F., Schettini, R.: Anomaly detection in nanofibrous materials by CNN-based self-similarity. Sensors 18(1), 209 (2018). ISSN: 1424-8220, https://www.mdpi.com/1424-8220/18/1/209, https://doi.org/10.3390/s18010209
12. Manevitz, L.M., Yousef, M.: One-class SVMs for document classification. J. Mach. Learn. Res. 2, 139–154 (2002). ISSN: 1532-4435

13. Ruff, L., Vandermeulen, R., Goernitz, N., et al.: Deep one-class classification. In: Dy, J., Krause, A. (eds.) Proceedings of the 35th International Conference on Machine Learning, ser. Proceedings of Machine Learning Research, vol. 80, PMLR, October 2018, pp. 4393–4402 (2018). https://proceedings.mlr.press/v80/ruff18a.html

14. Wang, L., Zhang, D., Guo, J., Han, Y.: Image anomaly detection using normal data only by latent space resampling. Appl. Sci. **10**(23), 8660 (2020). ISSN: 2076-3417. https://www.mdpi.com/2076-3417/10/23/8660, https://doi.org/10.3390/app10238660

15. Baur, C., Wiestler, B., Albarqouni, S., Navab, N.: Deep autoencoding models for unsupervised anomaly segmentation in brain MR images. In: Crimi, A., Bakas, S., Kuijf, H., Keyvan, F., Reyes, M., van Walsum, T. (eds.) BrainLes 2018. LNCS, vol. 11383, pp. 161–169. Springer, Cham (2019). https://doi.org/10.1007/978-3-030-11723-8_16

16. Gong, D., Liu, L., Le, V., et al.: Memorizing normality to detect anomaly: memory-augmented deep autoencoder for unsupervised anomaly detection. arXiv: 1904.02639 [cs.CV] (2019)

17. Tan, J., Hou, B., Batten, J., Qiu, H., Kainz, B.: Detecting outliers with foreign patch interpolation. arXiv: 2011.04197 [cs.CV] (2020)

18. Ronneberger, O., Fischer, P., Brox, T.: U-net: convolutional networks for biomedical image segmentation. CoRR, vol. abs/1505.04597 arXiv: 1505.04597 (2015)

19. Zhou, Z., Siddiquee, M.M.R., Tajbakhsh, N., Liang, J.: UNet++: a nested u-net architecture for medical image segmentation. CoRR, vol. abs/1807.10165 arXiv: 1807.10165 (2018)

20. Su, R., Zhang, D., Liu, J., Cheng, C.: MSU-Net: multi-scale U-Net for 2D medical image segmentation. Front. Genetics **12**, 140 (2021). ISSN: 1664-8021. https://www.frontiersin.org/article/10.3389/fgene.2021, https://doi.org/10.3389/fgene.2021.639930

21. Çiçek, Ö., Abdulkadir, A., Lienkamp, S.S., Brox, T., Ronneberger, O.: 3D U-Net: learning dense volumetric segmentation from sparse annotation. CoRR, vol. abs/1606.06650 arXiv: 1606.06650 (2016)

22. Milletari, F., Navab, N., Ahmadi, S.: V-Net: fully convolutional neural networks for volumetric medical image segmentation. CoRR, vol. abs/1606.04797 arXiv: 1606.04797 (2016)

SS3D: Unsupervised Out-of-Distribution Detection and Localization for Medical Volumes

Lars Doorenbos[✉], Raphael Sznitman, and Pablo Márquez-Neila

AIMI, ARTORG Center, University of Bern, Bern, Switzerland
{lars.doorenbos,raphael.sznitman,pablo.marquez}@unibe.ch

Abstract. We present an extension of the self-supervised outlier detection (SSD) framework [12] to the three-dimensional case. We first apply contrastive learning on a network using a general dataset of two-dimensional slices randomly sampled from all the available training data. This network serves as a latent embedding encoder of the input images. We model the in-distribution latent density as a multivariate Gaussian, fitted to the embeddings of the training slices. At test time, each test sample is scored by summing the Mahalanobis distances from all its slices to the means of the learned Gaussians. While mainly meant as a sample-level method, this approach additionally enables coarse localization, scoring each voxel by the minimum Mahalanobis distance among the slices that contain it. On the sample-level task of the 2021 MICCAI Medical Out-of-Distribution Analysis Challenge [20], our method ranked second on the challenging abdominal dataset, and fourth overall. Moreover, we show that with pretrained features and the right choice of architecture, a further boost in performance can be gained.

Keywords: out-of-distribution detection · Mahalanobis distance · contrastive learning

1 Introduction

Deep learning models have reached, and at times even surpassed, expert-level performance on a number of medical tasks in controlled academic environments [7,17]. However, translating these results to the real world still faces a number of critical hurdles. One of these is the detection and handling, at inference time, of samples drawn from outside the training data distribution. When presented with these *out-of-distribution* (OOD) samples, deep learning models exhibit undefined behaviour, and crucially, are unable to detect these cases by themselves. Hence, in order to be able to trust a deployed model, we should have a manner to detect and discard samples that it cannot reliably process. This is accomplished through the addition of an *OOD detector*, which ensures only samples drawn from the training distribution pass through for evaluation to the underlying model.

© Springer Nature Switzerland AG 2022
M. Aubreville et al. (Eds.): MIDOG 2021/MOOD 2021/L2R 2021, LNCS 13166, pp. 111–118, 2022.
https://doi.org/10.1007/978-3-030-97281-3_17

While some OOD detectors assume the availability of a large set of annotated training examples, these will not always be available. Consequently, they should be able to function in an unsupervised manner, learning the training distribution without labels and preventing any input that deviates too far from it from passing through.

For two-dimensional images, a wide variety of these unsupervised OOD detectors is available in the literature. They are typically based on frameworks such as generative models [3,11,13], autoencoders [6,19] or other types of representation learning [1,4,5], with most recently a trend towards contrastive learning [10,12,14,15,18].

The 3-dimensional case, on the other hand, has received far less attention. Current state-of-the-art methods in the medical domain operate along one plane, and aggregate the scores over slices to obtain a final volume-level result [9,16]. By committing to one plane, however, they might be biased towards a certain orientation of anomaly.

Our proposed method, an extension of the self-supervised outlier detection (SSD) [12] framework for volumetric data, overcomes this issue by combining results from all three anatomical planes. We submitted our approach to the sample-level task of the MICCAI Medical Out-of-Distribution Analysis Challenge (MOOD) [20], where it placed second on the challenging abdominal task, and fourth overall. Nevertheless, the MICCAI challenge did not allow methods to use pre-trained features from general computer vision datasets, such as ImageNet. By removing this constraint, we demonstrate that much better results can be obtained without any neural network training or fine-tuning of hyperparameters.

2 Approach

Our method relies on a neural network to extract latent embeddings, which are subsequently used to model the latent data distribution. A typical strategy is to choose a network pre-trained on ImageNet as this latent embedding encoder. However, given that the MOOD challenge did not allow the use of networks pre-trained on external datasets, our submitted approach makes use of self-supervised contrastive learning using exclusively the MOOD challenge datasets.

2.1 Datasets

The MOOD challenge provided two datasets: **Brain,** 800 MRI scans of size $256 \times 256 \times 256$ manually verified to hold no anomalies; and **Abdominal,** 550 anomaly-free CT scans of size $512 \times 512 \times 512$. Each dataset includes four toy test cases, which we use for validation.

2.2 Contrastive Learning

To perform contrastive learning, we first generated a temporary dataset by combining all the available training images from the challenge. In particular, we

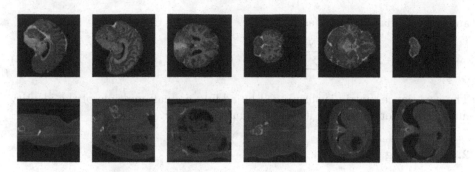

Fig. 1. Sample images from our 2D dataset used for contrastive learning.

randomly sampled slices from the coronal, sagittal and axial planes from both datasets where at least 1% of the pixels were non-zero. This resulted in a total of 81563 images (Fig. 1), that were used exclusively to solve the contrastive learning problem.

Our self-supervised learning approach of choice was the Simple framework for Contrastive Learning of visual Representations (SimCLR) [2]. SimCLR applies two stochastic augmentations to an image and projects these to a latent space. It then optimizes the network by minimizing the latent distance between the two versions of the same image while maximizing the distances to the other samples in the batch. This is done by minimizing the contrastive loss for minibatches of size N,

$$\ell_{i,j} = -\log \frac{\exp(\text{sim}(z_i, z_j)/\tau)}{\sum_{k=1}^{2N} \mathbb{1}_{[k \neq i]} \exp(\text{sim}(z_i, z_k)/\tau)}, \tag{1}$$

for augmentations i and j, where z represents the latent vector, τ the temperature, and $\text{sim}(x, y)$ the cosine similarity. We used the resulting network as our latent embedding encoder in all our experiments.

2.3   Training

Training consists of fitting multivariate Gaussians (MVG) to the distribution of the latent embeddings of the training data $\{x_i\}_{i=1}^N$. This is done independently for each anatomical plane —coronal, sagittal, axial— and slice. In particular, let $\{x_{i,s,p}\}_{i=1}^N$ be the set of two-dimensional slices of the training data at slice location s for a given anatomical plane $p \in P = \{\text{coronal, sagittal, axial}\}$. We extract the latent embeddings by average pooling the feature maps at the penultimate layer of our network, which gives a collection of descriptors $\{f(x_{i,s,p})\}_{i=1}^N$. We fit a MVG to this set of descriptors by computing the mean as

$$\mu_{p,s} = \frac{1}{N} \sum_{i=1}^N f(x_{i,p,s}), \tag{2}$$

and the covariance matrix as

$$\Sigma_{p,s} = \frac{1}{N} \sum_{i=1}^{N} (f(\mathbf{x}_{i,p,s}) - \boldsymbol{\mu}_{p,s})(f(\mathbf{x}_{i,p,s}) - \boldsymbol{\mu}_{p,s})^{\mathsf{T}}. \tag{3}$$

The covariance matrix is shrinkaged by adding a small multiple of the identity matrix in order to prevent it from becoming singular.

2.4 Scoring Slices

At inference time, the score at slice s for plane p is given by the Mahalanobis distance [8] between its latent embedding and the mean of the corresponding MVG,

$$M_{p,s}(\mathbf{x}) = d_{\Sigma_{p,s}}(f_{p,s}(\mathbf{x}), \boldsymbol{\mu}_{p,s}) = \sqrt{(f_{p,s}(\mathbf{x}) - \boldsymbol{\mu}_{p,s})^{\mathsf{T}} \Sigma_{p,s}^{-1} (f_{p,s}(\mathbf{x}) - \boldsymbol{\mu}_{p,s})}. \tag{4}$$

We found that slices at the edges of the brain were scoring disproportionally high (Fig. 2). To combat this, we ignore the anomaly scores of slices where more than 95% of pixels are zero.

2.5 Combining Dimensions

The final sample level score is given by the sum over all slices from the three planes,

$$S(x) = \sum_{p \in P} \sum_{s=1}^{d} M_{p,s}(\mathbf{x}). \tag{5}$$

With our approach, a coarse localization is possible. For a voxel with coordinates (i, j, z), the score is given by

$$S_{i,j,z}(x) = \min(M_{\text{coronal},i}(\mathbf{x}), M_{\text{sagittal},j}(\mathbf{x}), M_{\text{axial},z}(\mathbf{x})). \tag{6}$$

We found that smoothing the Mahalanobis distances along slices increases the localization performance. While this approach is only able to segment boxes, pointing the user towards the location of the anomaly should often suffice in practice.

3 Experiments

When evaluated on the MOOD toy data, which only contains four volumes per modality, our method reaches perfect accuracy at the sample level. Hence, in order to be able to differentiate between the performance of our models, we added a third evaluation alongside the pixel-level task, where we compute the slice-wise performance. A slice is considered OOD when it contains at least one OOD-labelled pixel.

For these three evaluations we show an ablation study in Tables 1 and 2. Using the self-supervised training over random weights results in the largest

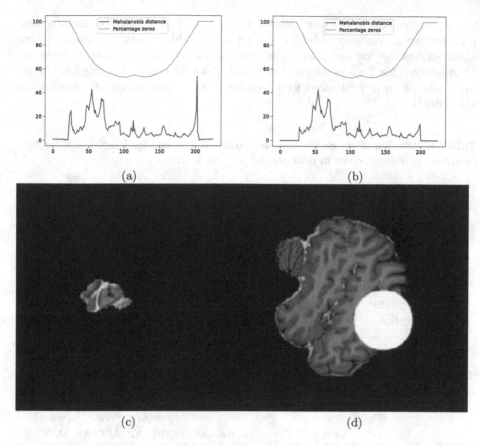

Fig. 2. Mahalanobis distance per slice along the sagittal plane for toy brain sample #4 from the MOOD challenge. (a) shows the two spurious peaks at the edges of the brain, (b) shows our correction. (c) shows the most outlying slice before the correction, (d) after.

gain in performance. The correction for slices where less than 5% of pixels are non-zero and the smoothing of the slice-wise responses further improve results.

Interestingly, our method is unable to perform well on the abdominal data when evaluated slice-wise, despite a perfect sample-level score on the toy data, and a second place in the MOOD challenge. We find this to be a result of OOD regions not being scored higher than some natural variations, see Fig. 3. The fact that our method reached the second place in the MOOD challenge on this dataset suggests that all current methods struggle with complex cases like these.

If we relax the constraint of not using external datasets to train the latent embedding network, we can boost the performance even further. Tables 1 and 2 also show results when using networks pre-trained on ImageNet instead of contrastive learning of the challenge datasets. With a ResNet-152 architecture, our approach is able to improve upon these features on the brain dataset, while being worse on the abdominal case. However, with an EfficientNet-b0 architecture, performance is consistently better in both cases. Additionally, the smaller latent space makes the method faster, and the contrastive learning step is bypassed completely. This finding suggests the field of OOD might be misguided in its over-reliance on using ResNet as a baseline model, upon which all comparisons are built [1,5,10,12,14,15,18].

Table 1. Ablation study on the toy brain data. RN stands for ResNet, EN for EfficientNet. Best result given in bold, second best underlined.

	Features	Correction	Smoothing	Sample AUC	Sample AP	Slice AUC	Slice AP	Pixel AUC	Pixel AP
Random weights	Random	✗	✗	100	100	74	36	84	8
+ contrastive learning	Contrastive	✗	✗	100	100	84	64	88	27
+ edge correction	Contrastive	✓	✗	100	100	84	72	90	34
+ smoothing	Contrastive	✗	✓	100	100	84	68	88	31
SS3D	Contrastive	✓	✓	100	100	<u>88</u>	<u>80</u>	90	<u>38</u>
Pretrained RN-152	Pretrained	✓	✓	100	100	86	77	**94**	29
Pretrained EN-b0	Pretrained	✓	✓	100	100	**92**	**87**	<u>93</u>	**63**

Table 2. Ablation study on the toy abdominal data. RN stands for ResNet, EN for EfficientNet. Best result given in bold, second best underlined.

	Features	Correction	Smoothing	Sample AUC	Sample AP	Slice AUC	Slice AP	Pixel AUC	Pixel AP
Random weights	Random	✗	✗	100	100	48	16	63	1
+ contrastive learning	Contrastive	✗	✗	100	100	49	16	66	1
+ edge correction	Contrastive	✓	✗	100	100	50	20	68	1
+ smoothing	Contrastive	✗	✓	100	100	52	17	66	1
SS3D	Contrastive	✓	✓	100	100	53	19	69	1
Pretrained RN-152	Pretrained	✓	✓	100	100	<u>80</u>	<u>56</u>	<u>91</u>	<u>13</u>
Pretrained EN-b0	Pretrained	✓	✓	100	100	**86**	**70**	**94**	**21**

In Fig. 4 we show how the localization is correctly able to detect an abnormal region in a volume from the MOOD brain test set. While not segmenting the spherical lesion fully, the end-user would be successfully pointed towards the anomalous region. Additionally, inference with our method is very fast, with all 667 brain test samples predicted in only 124 s on the MOOD submission platform, and can thus be used in real-time.

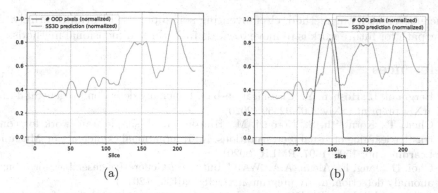

(a)   (b)

Fig. 3. (a) shows the predictions along slices for a normal sample, (b) the predictions with an anomalous region centered around slice 95.

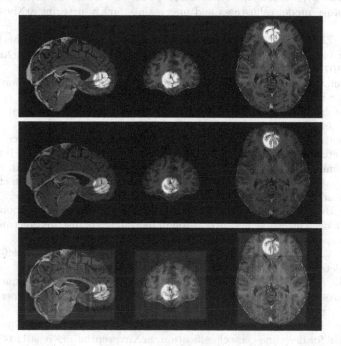

Fig. 4. Image, ground truth and localization by our model for a volume from the MOOD test set.

4   Conclusion

We presented an extension of the SSD framework [12] for use with medical volumes. Besides sample-level OOD detection, coarse localization is made possible through the combination of results along all three planes. For any real-world application, unconstrained by challenge restrictions, we recommend using a pretrained

EfficientNet with our method, circumventing self-supervised learning entirely. A clear path for future work is to move towards fully three-dimensional methods.

References

1. Bergman, L., Hoshen, Y.: Classification-based anomaly detection for general data. arXiv preprint arXiv:2005.02359 (2020)
2. Chen, T., Kornblith, S., Norouzi, M., Hinton, G.: A simple framework for contrastive learning of visual representations. In: International Conference on Machine Learning, pp. 1597–1607. PMLR (2020)
3. Choi, H., Jang, E., Alemi, A.A.: WAIC, but why? Generative ensembles for robust anomaly detection. arXiv preprint arXiv:1810.01392 (2018)
4. Golan, I., El-Yaniv, R.: Deep anomaly detection using geometric transformations. In: Advances in Neural Information Processing Systems, pp. 9758–9769 (2018)
5. Hendrycks, D., Mazeika, M., Kadavath, S., Song, D.: Using self-supervised learning can improve model robustness and uncertainty. arXiv preprint arXiv:1906.12340 (2019)
6. Hou, J., Zhang, Y., Zhong, Q., Xie, D., Pu, S., Zhou, H.: Divide-and-assemble: learning block-wise memory for unsupervised anomaly detection. arXiv preprint arXiv:2107.13118 (2021)
7. Kurmann, T., et al.: Expert-level automated biomarker identification in optical coherence tomography scans. Sci. Rep. $9(1)$, 1–9 (2019)
8. Mahalanobis, P.C.: On the generalized distance in statistics. National Institute of Science of India (1936)
9. Marimont, S.N., Tarroni, G.: Anomaly detection through latent space restoration using vector quantized variational autoencoders. In: 2021 IEEE 18th International Symposium on Biomedical Imaging (ISBI), pp. 1764–1767. IEEE (2021)
10. Reiss, T., Hoshen, Y.: Mean-shifted contrastive loss for anomaly detection. arXiv preprint arXiv:2106.03844 (2021)
11. Schirrmeister, R.T., Zhou, Y., Ball, T., Zhang, D.: Understanding anomaly detection with deep invertible networks through hierarchies of distributions and features. arXiv preprint arXiv:2006.10848 (2020)
12. Sehwag, V., Chiang, M., Mittal, P.: SSD: a unified framework for self-supervised outlier detection. arXiv preprint arXiv:2103.12051 (2021)
13. Serrà, J., Álvarez, D., Gómez, V., Slizovskaia, O., Núñez, J.F., Luque, J.: Input complexity and out-of-distribution detection with likelihood-based generative models. arXiv preprint arXiv:1909.11480 (2019)
14. Sohn, K., Li, C.L., Yoon, J., Jin, M., Pfister, T.: Learning and evaluating representations for deep one-class classification. arXiv preprint arXiv:2011.02578 (2020)
15. Tack, J., Mo, S., Jeong, J., Shin, J.: CSI: novelty detection via contrastive learning on distributionally shifted instances. arXiv preprint arXiv:2007.08176 (2020)
16. Tan, J., Hou, B., Batten, J., Qiu, H., Kainz, B.: Detecting outliers with foreign patch interpolation. arXiv preprint arXiv:2011.04197 (2020)
17. Tang, Y.X., et al.: Automated abnormality classification of chest radiographs using deep convolutional neural networks. NPJ Digital Medicine $3(1)$, 1–8 (2020)
18. Xiao, Z., Yan, Q., Amit, Y.: Do we really need to learn representations from in-domain data for outlier detection? arXiv preprint arXiv:2105.09270 (2021)
19. Yoon, S., Noh, Y.K., Park, F.C.: Autoencoding under normalization constraints. arXiv preprint arXiv:2105.05735 (2021)
20. Zimmerer, D., et al.: Medical out-of-distribution analysis challenge 2021, March 2021. https://doi.org/10.5281/zenodo.4573948

MetaDetector: Detecting Outliers by Learning to Learn from Self-supervision

Jeremy Tan[1(✉)], Turkay Kart[1], Benjamin Hou[1], James Batten[1],
and Bernhard Kainz[1,2]

[1] Imperial College London, London SW7 2AZ, UK
j.tan17@imperial.ac.uk
[2] Friedrich–Alexander University Erlangen–Nürnberg, Erlangen, Germany

Abstract. Using self-supervision in anomaly detection can increase sensitivity to subtle irregularities. However, increasing sensitivity to certain classes of outliers could result in decreased sensitivity to other types. While a single model may have limited coverage, an adaptive method could help detect a broader range of outliers. Our proposed method explores whether meta learning can increase the adaptability of self-supervised methods. Meta learning is often employed in few-shot settings with labelled examples. To use it for anomaly detection, where labelled support data is usually not available, we instead construct a self-supervised task using the test input itself and reference samples from the normal training data. Specifically, patches from the test image are introduced into normal reference images. This forms the basis of the few-shot task. During training, the same few-shot process is used, but the test/query image is substituted with a normal training image that contains a synthetic irregularity. Meta learning is then used to learn how to learn from the few-shot task by computing second order gradients. Given the importance of screening applications, e.g. in healthcare or security, any adaptability in the method must be counterbalanced with robustness. As such, we add strong regularization by i) restricting meta learning to only layers near the bottleneck of our encoder-decoder architecture and ii) computing the loss at multiple points during the few-shot process.

Keywords: Outlier detection · Self-supervised learning · Meta-learning

1 Introduction

The main goal in outlier detection is to detect unanticipated irregularities. Many important problems can be framed in this way, including content moderation, security, and disease screening. All of these tasks can be very taxing on workers and phenomena such as inattentional blindness can make it particularly difficult to detect unexpected stimuli [6]. Machine learning methods have the potential to assist in many of these tasks. While most methods require labelled data to achieve expert-level performance, e.g. supervised methods for detecting breast

© Springer Nature Switzerland AG 2022
M. Aubreville et al. (Eds.): MIDOG 2021/MOOD 2021/L2R 2021, LNCS 13166, pp. 119–126, 2022.
https://doi.org/10.1007/978-3-030-97281-3_18

cancer [22] or retinal disease [4], self-supervised methods have recently begun to close the gap [3,10,11,16].

Similar self-supervised approaches also exist in outlier detection [8,19]. However, applications such as medical imaging can require very specific and subtle features that are difficult to learn in an unsupervised manner. Many diseases can only be detected by those with domain expertise; but being a specialist does not always improve detection. The "cost of expertise" is a bias toward familiar patterns, and can cause rigidity in perception of new stimuli. For example, one study found that physicians have a tendency to make diagnoses related to their speciality, even when examining cases outside of their domain [9]. This is especially problematic in open-ended problems such as outlier detection because there are no restrictions on what pathologies may appear.

As such, anomalous features can be subtle and disease specific, but can also vary immensely across pathologies. Recent methods have increased sensitivity to subtle outliers through self-supervision [20,21]. Our aim in this work is to explore whether meta learning can improve the adaptability of these methods. We construct a self-supervised few-shot task to be used during both training and testing. During training, we use second order gradients [7] to optimize for initial parameters that are adaptable to different tasks. In testing, the few-shot task gives the model a chance to adapt to features in the test data, potentially priming the model for better detection.

2   Related Work

Many methods have been developed to tackle outlier detection from different perspectives. Reconstruction-based methods use auto-encoders [2] or generative models [18] to reproduce or restore [15] the normal components of the image. Errors in the reconstruction are used to highlight abnormalities. This is most effective for abnormalities that exhibit large intensity differences. Self-supervised methods are trained on proxy tasks that can either exploit (i) whole image augmentations that help the network to learn holistic features and major landmarks in the normal data [8,19] or (ii) patch-based augmentations that increase sensitivity to sub-image anomalies [14,20,21]. The design of the self-supervised task can influence which types of features are learned and consequently which types of anomalies are detected. This can help increase sensitivity to specific types of irregularities, but it can also limit detection of other types of anomalies.

Meta learning can be applied to few-shot problems to allow models to adapt quickly to new tasks. One of the most ubiquitous strategies in this area is model-agnostic meta-learning (MAML) [7]. While first order gradients point toward parameters that give better outputs, second order gradients point toward parameters that give better few-shot gradients. By backpropagating through the few-shot optimization steps, MAML aims to learn an initialization point that benefits the most from the few-shot gradients. This strategy is typically used in settings with labelled examples, but there are also extensions in unsupervised settings that use clustering [12]. There are even few-shot methods in outlier detection;

however, these exploit a small amount of real anomalous data [5], or use multi-class data organized into normal and anomalous categories [13]. The goal of this work is to explore meta learning in a setting where the few-shot task is self-supervised.

3 Method

Our proposed method, MetaDetector, is a meta learning approach for self-supervised outlier detection. In this section we briefly describe the self-supervised tasks, the meta learning process, and the regularization involved in training.

The two self-supervised tasks used in this method are foreign patch inter-polation (FPI) [20] and Poisson image interpolation (PII) [21]. In both tasks, a random patch is taken from one image and introduced into another image. This is done through linear interpolation in the case of FPI and by Poisson image editing [17] in the case of PII. In both cases, the corresponding label is a mask of the altered patch that is scaled by the blending factor. Figure 1 depicts an example of FPI, where a patch in image x_i has been altered to produce $\tilde{x}_i$. Using $\tilde{x}_i$ as an input, the output of the network, $A_s(\tilde{x}_i)$, is compared with the label $\tilde{y}_i$ using a binary cross-entropy loss.

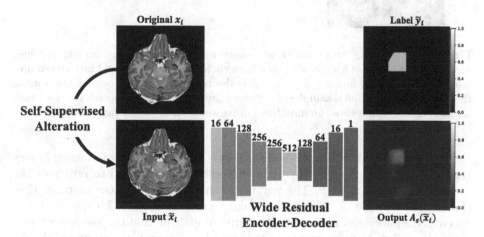

Fig. 1. Network architecture with example inputs, labels, and outputs. Quantities above each residual block indicate the number of feature channels.

The meta learning component of our method is based on MAML [7] and it involves an inner and outer optimization loop. The inner loop is the few-shot optimization process. It involves a query image x_q, which may or may not be anomalous, and a reference image x_r which is a normal image from the training data. Random patches from the query image are introduced into the reference image using FPI, producing $\tilde{x}_r$. This creates an input and label, as shown in Fig. 1, which can be used to take one optimization step in the inner loop. This can be repeated k times depending on how many steps are desired in the few-shot process. In each step, the query image remains the same, but a new reference image and a new random patch are selected. In all of our experiments we use $k = 2$. This few-shot process is applied during testing and training. In testing, the query image is an unknown sample, x_q, but during training the query image is a normal sample (from the training data) that has been altered with FPI or PII, i.e. $\widetilde{x}_q$.

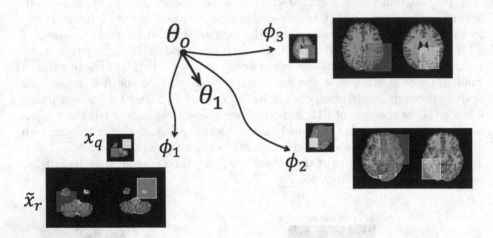

Fig. 2. Meta learning process with self-supervised few-shot task. Random patches, highlighted in blue and yellow, are taken from each query image x_q and introduced into reference images, x_r, forming $\widetilde{x}_r$. Each ϕ is the result of few-shot optimization using these synthetically altered samples, $\widetilde{x}_r$. Second order gradients from each few-shot task are aggregated to improve the initialization parameters, θ. (Color figure online)

The loss for both self-supervised tasks is a pixel-wise regression using binary cross-entropy, as defined by Eq. 1. In this equation, f_θ is used to represent the model parameterized by θ. The parameter updates in the inner loop are thus characterized by Eq. 2. This is a standard gradient update. Meanwhile Eq. 3 specifies the parameter updates for the outer loop. Note that the loss is evaluated using the updated parameters from the inner loop, ϕ, and the query sample, $\widetilde{x}_q$. However, the gradient is taken with respect to the initial parameters, θ, which means that gradient must flow through the gradient steps of the inner loop.

$$\mathcal{L}_{\text{bce}}(\widetilde{x}_r, \widetilde{y}_r, f_\theta) = -\widetilde{y}_r log f_\theta(\widetilde{x}_r) - (1 - \widetilde{y}_r)\log(1 - f_\theta(\widetilde{x}_r)) \tag{1}$$

$$\phi_i = \theta - \alpha \nabla_\theta \mathcal{L}_{\text{bce}}(\widetilde{x}_r, \widetilde{y}_r, f_\theta) \tag{2}$$

$$\theta_i = \theta - \beta \nabla_\theta \mathcal{L}_{\text{bce}}(\widetilde{x}_q, \widetilde{y}_q, f_\phi) \tag{3}$$

Note that Eqs. 2 and 3 are simplified for legibility. In reality, ϕ_i is the result of several gradient steps, and θ_i is updated using an aggregate of multiple inner loops. This is depicted in Fig. 2. In our experiments, the step sizes α and β are set to 1×10^{-2} and 1×10^{-3}, respectively.

The last component of our method is the regularization. Since the model is being trained for fast adaptation, the predictions from the model could change drastically after only a few gradient steps. This is useful for adaptability, but it could also negatively impact robustness. We add strong regularization to the optimization process in two ways. First, we restrict meta learning to parameters in the bottleneck residual block (orange in Fig. 1). The rest of the parameters are learned using standard self-supervised training (as in FPI [20] and PII [21]) using only the query samples, $\tilde{x}_q$. This limits the number of parameters that can change during few-shot adaptation and drastically reduces computational costs.

Our second form of regularization is a multi-step loss based on a strategy proposed in MAML++ [1]. The original MAML method computes outer loop gradients according to Eq. 3, specifically using f_ϕ, the parameters that are reached after taking the final few-shot step. With a multi-step approach, we compute the loss after every few-shot step [1]. The total loss is a weighted sum, where coefficients increase linearly with step number.

To run MetaDetector in evaluation mode, only the few-shot process is performed. Patches from the test image, x_q, are introduced into normal reference images, x_r, which creates self-supervised samples $\tilde{x}_r$. The model is trained on these samples to reach the adapted parameters, ϕ. The updated model is then used to run inference on the original test image, x_q. The output from this inference is used directly as an anomaly score map, $A_s(x_q)$.

4   Evaluation and Results

We train MetaDetector on normal brain MRI and abdominal CT data from the medical out-of-distribution (MOOD) analysis challenge [23]. Since this dataset only includes normal samples, we evaluate on a synthetic test dataset that includes spheres with uniform intensity shifts, noise additions, sink/source deformations, uniform translations, and reflections across axes of symmetry. The details and the code to reproduce these test cases are provided in Tan et al. [20].

Table 1. Average precision for brain and abdominal synthetic test data [20] (originally from the MOOD challenge [23]).

Anatomy	Method	Subject-level AP	Pixel-level AP
Brain	FPI [20]	0.9723	0.7319
	MetaDetector	**0.9989**	**0.8551**
Abdomen	FPI [20]	0.8854	**0.6229**
	MetaDetector	**0.9694**	0.1657

Preliminary results are given in Table 1 and Fig. 3. Overall, MetaDetector outperforms FPI [20] on synthetic test data. The low abdominal pixel-level score is likely due to a bias in the synthetic test data toward visibility in the coronal view. While FPI [20] was trained using 2D coronal slices [20], MetaDetector uses transverse slices at half the resolution (256×256) in order to meet computational restrictions. Qualitative examples in Fig. 3 (provided by the MOOD organizers [23]) indicate that MetaDetector can localize different types of subtle abnormalities. However, false positives appear to be quite common. This could be either due to (i) synthetic training samples that are too subtle or (ii) the few-shot process priming the model to recognize healthy tissue from the query image as abnormal.

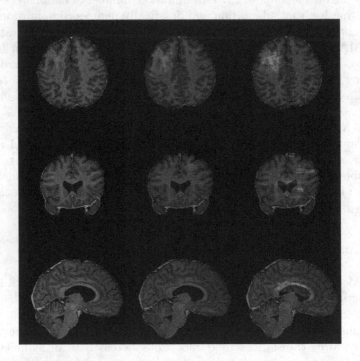

Fig. 3. Qualitative samples from the MOOD challenge [23]. Input, ground truth, and prediction (from left to right).

5   Discussion and Conclusion

We explore the use of meta learning in self-supervised outlier detection. Meta learning techniques such as MAML [7] can help models to quickly adapt to new data. We present an approach that uses self-supervision for the few-shot task, which allows training with only normal data. Early results indicate that the proposed method can outperform other self-supervised methods, such as FPI [20], on synthetic test data. This basic framework also has the potential to make outlier detection more adaptive, to better handle new stimuli. In future work, we aim to design more varied self-supervised tasks to encourage the model to rely more on information gained in the few-shot learning process.

Acknowledgements. JT was supported by an Imperial College London President's Scholarship. JB was supported by the UKRI CDT in AI for Healthcare http:// ai4health.io (Grant No. P/S023283/1). This work was supported by the London Medical Imaging & AI Centre for Value Based Healthcare (104691), EP/S013687/1, EP/R005982/1 and Nvidia for the ongoing donations of high-end GPUs.

References

1. Antoniou, A., Edwards, H., Storkey, A.: How to train your MAML. In: International Conference on Learning Representations (2018)
2. Baur, C., Denner, S., Wiestler, B., Navab, N., Albarqouni, S.: Autoencoders for unsupervised anomaly segmentation in brain MR images: a comparative study. In: Medical Image Analysis, p. 101952 (2021)
3. Chen, T., Kornblith, S., Norouzi, M., Hinton, G.: A simple framework for contrastive learning of visual representations. In: International Conference on Machine Learning, pp. 1597–1607. PMLR (2020)
4. De Fauw, J., et al.: Clinically applicable deep learning for diagnosis and referral in retinal disease. Nat. Med. **24**(9), 1342–1350 (2018)
5. Ding, K., Zhou, Q., Tong, H., Liu, H.: Few-shot network anomaly detection via cross-network meta-learning. In: Proceedings of the Web Conference 2021, pp. 2448–2456 (2021)
6. Drew, T., Võ, M., Wolfe, J.: The invisible gorilla strikes again: sustained inattentional blindness in expert observers. Psychol. Sci. **24**(9), 1848–1853 (2013)
7. Finn, C., Abbeel, P., Levine, S.: Model-agnostic meta-learning for fast adaptation of deep networks. In: International Conference on Machine Learning, pp. 1126–1135. PMLR (2017)
8. Golan, I., El-Yaniv, R.: Deep anomaly detection using geometric transformations. In: Advances in Neural Information Processing Systems, pp. 9758–9769 (2018)
9. Hashem, A., Chi, M.T., Friedman, C.P.: Medical errors as a result of specialization. J. Biomed. Inform. **36**(1–2), 61–69 (2003)
10. He, K., Fan, H., Wu, Y., Xie, S., Girshick, R.: Momentum contrast for unsupervised visual representation learning. In: Proceedings of the IEEE/CVF Conference on Computer Vision and Pattern Recognition, pp. 9729–9738 (2020)
11. Hénaff, O.J., et al.: Data-efficient image recognition with contrastive predictive coding. arXiv preprint arXiv:1905.09272 (2019)
12. Hsu, K., Levine, S., Finn, C.: Unsupervised learning via meta-learning. In: ICLR (2019)
13. Jeong, T., Kim, H.: OOD-MAML: meta-learning for few-shot out-of-distribution detection and classification. In: Advances in Neural Information Processing Systems, vol. 33 (2020)
14. Li, C.L., Sohn, K., Yoon, J., Pfister, T.: CutPaste: self-supervised learning for anomaly detection and localization. In: Proceedings of the IEEE/CVF Conference on Computer Vision and Pattern Recognition, pp. 9664–9674 (2021)
15. Marimont, S.N., Tarroni, G.: Anomaly detection through latent space restoration using vector quantized variational autoencoders. In: 2021 IEEE 18th International Symposium on Biomedical Imaging (ISBI), pp. 1764–1767. IEEE (2021)

16. van den Oord, A., Li, Y., Vinyals, O.: Representation learning with contrastive predictive coding. arXiv preprint arXiv:1807.03748 (2018)
17. Pérez, P., Gangnet, M., Blake, A.: Poisson image editing. In: ACM SIGGRAPH 2003 Papers, pp. 313–318 (2003)
18. Schlegl, T., Seeböck, P., Waldstein, S.M., Langs, G., Schmidt-Erfurth, U.: f-AnoGAN: fast unsupervised anomaly detection with generative adversarial networks. Med. Image Anal. **54**, 30–44 (2019)
19. Tack, J., Mo, S., Jeong, J., Shin, J.: CSI: novelty detection via contrastive learning on distributionally shifted instances. arXiv preprint arXiv:2007.08176 (2020)
20. Tan, J., Hou, B., Batten, J., Qiu, H., Kainz, B.: Detecting outliers with foreign patch interpolation. arXiv preprint arXiv:2011.04197 (2020)
21. Tan, J., Hou, B., Day, T., Simpson, J., Rueckert, D., Kainz, B.: Detecting outliers with Poisson image interpolation. In: de Bruijne, M., et al. (eds.) MICCAI 2021. LNCS, vol. 12905, pp. 581–591. Springer, Cham (2021). https://doi.org/10.1007/978-3-030-87240-3_56
22. Wu, N., et al.: Deep neural networks improve radiologists' performance in breast cancer screening. IEEE Trans. Med. Imaging **39**(4), 1184–1194 (2019)
23. Zimmerer, D., et al.: Medical out-of-distribution analysis challenge (2020)

AutoSeg - Steering the Inductive Biases for Automatic Pathology Segmentation

Felix Meissen[1,2(✉)], Georgios Kaissis[1,2,3], and Daniel Rueckert[1,2,3]

[1] Technical University of Munich (TUM), Munich, Germany
felix.meissen@tum.de
[2] Klinikum Rechts der Isar, Munich, Germany
[3] Imperial College London, London, UK

Abstract. In medical imaging, un-, semi-, or self-supervised pathology detection is often approached with anomaly- or out-of-distribution detection methods, whose inductive biases are not intentionally directed towards detecting pathologies, and are therefore sub-optimal for this task. To tackle this problem, we propose AutoSeg, an engine that can generate diverse artificial anomalies that resemble the properties of real-world pathologies. Our method can accurately segment unseen artificial anomalies and outperforms existing methods for pathology detection on a challenging real-world dataset of Chest X-ray images. We experimentally evaluate our method on the Medical Out-of-Distribution Analysis Challenge 2021 (Code available under: https://github.com/FeliMe/autoseg).

Keywords: Self-Supervised Anomaly Segmentation · Anomaly Detection · Inductive Bias

1 Introduction

Anomalies are samples that deviate from a predefined norm. In medical images, these can manifest in various ways. Inaccuracies in image acquisition – like motion artifacts in MRI – can be considered as anomalies, as well as pathologies like tumors, natural intra-patient variations, or images from other modalities, such as natural images. In most applications however, we are interested in detecting pathologies. Therefore, the problem is ill-defined, and detection methods that follow this broad definition are likely to struggle with this difficult task. Moreover, all anomaly detection models have some form of inductive bias, making certain types of anomalies harder to detect than others. Especially reconstruction-based anomaly detection methods have a strong inductive bias because of the scoring function that is based on relative intensity differences between the original image and the reconstruction. However, the inductive biases of many methods are not steered towards that problem. Recent works have shown that machine learning models can successfully be trained on synthetic data [12,16]. Apart from performance improvements, this allows for more

---

G. Kaissis and D. Rueckert—Equal contribution.

M. Aubreville et al. (Eds.): MIDOG 2021/MOOD 2021/L2R 2021, LNCS 13166, pp. 127–135, 2022.
https://doi.org/10.1007/978-3-030-97281-3_19

control over the models' detection characteristics. We therefore, present a new approach that creates a useful inductive bias to better detect pathologies in medical images.

Our contributions are the following:

- We propose AutoSeg, a novel strategy for generating artificial anomalies that represent characteristics of real-world anomalies better than existing methods.
- We evaluate our approach on multiple modalities including a challenging and unsolved real-world dataset.
- We achieve state of the art detection performance on artificial and real-world anomalies.

2 Related Work

We sort related work in medical anomaly detection into three categories:

Reconstruction-based methods train a generative model – such as an Autoencoder (AE) or a Generative Adversarial Network (GAN) – on images from healthy subjects only. This way, the model learns the underlying distribution and will fail to reconstruct regions in images that are anomalous, and were thus not observed during training. Prominent representatives of these methods are [1,10]. In [1], the authors use a convolutional Autoencoder as their generative model. In [10], Schlegl et al. train a GAN to represent the "healthy" distribution and during test-time use restoration to find an image that is both close to the input image and the learned manifold. Recently, Baur et al. [2] compared all reconstruction-based methods in a large study. We refer the reader to their work for a more thorough overview thereof. This family of methods, however, has a very strong inductive bias, and it was recently shown by Meissen et al. [5] that for brain MRI, it can be outperformed via simple thresholding.

The second category contains methods that attempt to directly evaluate the likelihood of a sample being from the "healthy" distribution. They also use generative models trained on images from healthy patients only to compute the likelihood. Pinaya et al. [9] train an ensemble of autoregressive transformers on the latent space of a pretrained fully-convolutional Vector Quantised-Variational Autoencoder (VQ-VAE) to estimate the likelihood of every spatial feature in this latent space. Zimmerer et al. [17] use the gradient of the evidence lower bound (ELBO) of a trained Variational Autoencoder (VAE) to detect anomalies in brain MRI. The ELBO is a lower bound on the actual log-likelihood of a sample x: $\mathrm{ELBO}(x) \leq \log p(x)$.

Recently, training on synthetic data became a popular approach for machine learning in regimes where annotated data is hard to acquire. It has been successfully applied to optical flow estimation [12] and face-related computer vision [16]. In medical imaging, Tan et al. [13] trained a self-supervised segmentation model to detect artificial anomalies on brain MRI and abdomen CT. The anomalies are created by selecting two images from different patients and interpolating between them at randomly sampled rectangular patches with random interpolation factors.

3 Method

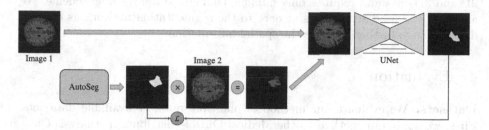

Fig. 1. Overview of our method. An anomaly mask is generated via AutoSeg. The anomalous texture is taken from a different image like in [13]. A UNet predicts the anomaly mask and strength of interpolation.

Baseline. We build our method upon the work of Tan *et al.* [13]. To create artificial anomalies, we choose two images at random, sample a rectangular mask at a random location inside the two, uniformly sample an interpolation factor in the interval $[0.05, 0.95]$, and interpolate the images at the sampled mask with the interpolation factor. A segmentation model is then trained to segment the anomalous region, as well as to predict the interpolation factor. Unlike the original work, we use a UNet [3] as our segmentation model.

AutoSeg. To increase the realism of the artificial anomalies, we propose AutoSeg, an engine to generate anomalies with diverse characteristics. AutoSeg is able to generate a specified number of anomalies in the shape of random polygons with a controllable number of vertices. This way, single large anomalies can be created – as it is common for tumors – or multiple smaller ones like in multiple sclerosis (MS) lesions. The artificial anomaly generation process consists of two parts: Creating the anomaly mask, and choosing the anomalous texture. We use our proposed AutoSeg for the former and adhere to the patch interpolation strategy of the baseline for the latter. Figure 1 shows an overview of our method.

Volumetric Data. Medical data is often three-dimensional. Despite that, most existing methods for anomaly detection are processing single slices independently, discarding all spatial information along one axis [1,2,9,10,13,17]. Using 3D convolutions is a potential solution to this problem but comes with significant computational costs. This forces the user to either downsample the data or to use a patch-based approach. Both alternatives limit the spatial resolution along all three axes. Our solution mimics how radiologists look at volumetric data and is similar to work of Perslev *et al.* [7]. Instead of providing only one viewing direction, we train our models with samples from all three viewing directions (axial, coronal, and sagittal) equally as in. During test-time, we perform inference on all three viewing directions and fuse the results by taking the average across all three predictions. This allows us to use the same architecture for 2D and 3D data and requires only minimal changes to the training scheme. We also feed $k = 3$ adjacent slices at once to the segmentation network as channel dimensions to further increase the spatial information.

4   Evaluation

Datasets. We evaluate our method on multiple publicly available data sets. First, we apply our method to the Medical Out-of-Distribution Analysis Challenge 2021 (MOOD) [8]. The data contains 800 brain T2 MRI scans of healthy young adults from the Human Connectome Project [14] with $256 \times 256 \times 256$ pixels per scan, and 550 abdominal CT scans of patients over 50 years of age [11] with $512 \times 512 \times 512$ pixels per scan. We split the two data sets into 60% training and 40% test data. From the training set, we use 5% to evaluate performance during training. Since no test data is publicly available for the MOOD challenge, we create our own artificial anomalies for evaluation on 75% of the held-out test set. For every anomalous sample A, we chose a sphere h in the scan at random and add one of eight anomaly types to the pixels in the sphere. We use the six anomaly types of Tan $et\ al.$ [13] and add two more, called local blur and slice shuffle. For the local blur, we create a blurred version A' of the input scan via Gaussian filtering and replace the original sample A with A' at the anomalous region h. For the slice shuffle, we choose a random axis and replace each slice along this axis in the patch with another slice inside that patch.

To test the suitability of our method for clinical applications, evaluation on a non-trivial real-world dataset is necessary. Therefore, we further evaluate our method on the publicly available ChestX-ray14 dataset [15], containing $112, 120$ frontal-view X-ray images of $30, 805$ patients with 14 disease labels. The dataset also includes the disease bounding boxes for 984 images. We only consider $43, 322$ images of patients over 18 and with posteroanterior view from which we use 75% for training and the rest for evaluation. While the original image resolution is 1024×2014, we bilinearly downsample all images to 256×256 – maintaining the aspect ratio – to be in line with the brain MR images. Because of the obvious domain shift, we processed the male and female patients separately. This dataset is challenging for pathology detection, because of the large intra-patient and intra-image variance in the patients' position, their anatomy, and the presence of external objects such as pacemakers.

Experimental Setup. We implement our method in the PyTorch [6] framework. For the UNet, we choose the implementation of Buda $et\ al.$ [3]. We train our method for 5 epochs using the AdamW optimizer [4] with the default parameters for the UNet, a batch size of 8, and a learning rate of 0.0001. For the

ChestX-ray14 dataset, we choose a larger batch size of 64, a smaller model with width = 16, and train for 10 epochs. We compare our method against a reconstruction-based VAE and report the average precision for every experiment. For reference, we also include the performance a randomly guessing model would achieve. AutoSeg is tuned to generate single anomalies with 10 vertices, cubic spline interpolation between the vertices, and sizes uniformly sampled between 10% and 50%, 20% and 60%, and 5% and 70% of the image size for brain MRI, abdominal CT, and Chest X-ray respectively. We deliberately performed only minimal hyperparameter tuning – varying only the model width and learning rate – to emphasize the contribution of our method over optimization strategy improvements.

5 Results

Table 1. Average precision of our and comparing methods on unseen artificial anomalies of the held-out test set.

Method	Brain MRI		Abdomen CT	
	Sample	Pixel	Sample	Pixel
Random	0.750	0.003	0.750	0.004
VAE	0.740	0.016	0.749	0.011
UNet (Baseline)	0.956	0.427	-	-
+AutoSeg	0.999	0.954	-	-
+2.5D Training	**1.000**	**0.974**	**0.988**	**0.953**

Here we show the results of the evaluation described in Sect. 4. In Table 1 the segmentation and detection performance of our models on the artificial anomalies from the held-out test set is presented. Using our proposed AutoSeg yields the strongest performance improvement, leading to almost perfect segmentation. The worst performance in this experiment is achieved by the VAE, not substantially exceeding random guessing.

In Table 2, we present a detailed evaluation of the different artificial anomaly types. While performance on the UNet baseline is greatly different between different anomaly types, the models trained with AutoSeg show comparable average precision scores among all types. Slice shuffle anomaly benefits the most from using 2.5D information during training. This is expected behavior, as it is the only anisotropic anomaly and is easier to detect from some viewing directions than from others. Although all other anomalies are isotropic, 2.5D training also helps in these cases. Additionally, we evaluate our final model trained with AutoSeg and 2.5D information on the MOOD 2021 challenge [8]. Here, the model achieves 6th place in the sample-level, and 4th in the pixel-level evaluation.

Table 2. Average precision for pixel-wise evaluation on MOOD brain MRI.

	VAE	UNet(Baseline)	+AutoSeg	+2.5D Training
Local Blur	0.012	0.915	0.997	**0.999**
Slice Shuffle	0.008	0.134	0.930	**0.990**
Noise Addition	0.015	0.124	0.908	**0.936**
Reflection	0.018	0.451	0.986	**0.989**
Sink Deformation	0.017	0.045	0.922	**0.958**
Source Deformation	0.016	0.550	0.944	**0.968**
Uniform Addition	0.022	0.699	**0.987**	0.986
Uniform Shift	0.021	0.603	0.961	**0.967**
Total	0.016	0.427	0.954	**0.974**

Table 3 only shows sample-wise results, as there are no anomaly segmentations included in the ChestX-ray14 dataset. Here, the performance of our model trained with AutoSeg is substantially lower compared to the artificial anomalies but still outperforms all competing methods.

Table 3. Sample-wise average precision of our and comparing methods on posteroanterior images of patients over 18 from the ChestX-ray14 dataset, split by gender.

	Male	Female
Random	0.561	0.579
VAE	0.607	0.624
UNet(Baseline)	0.539	0.576
+AutoSeg	**0.643**	**0.647**

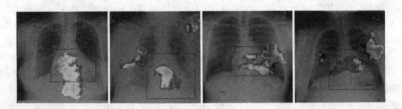

Fig. 2. Selected examples of patients with cardiomegaly overlaid with the predicted anomaly map and the ground truth bounding box.

Figure 3 displays the predicted anomaly maps and the ground truth segmentation maps of two random samples showing a "uniform shift" and a "noise addition" anomaly. The predictions are almost perfect, showing only some minor inaccuracies at the borders. Another slight error is visible in the random shift

Predictions

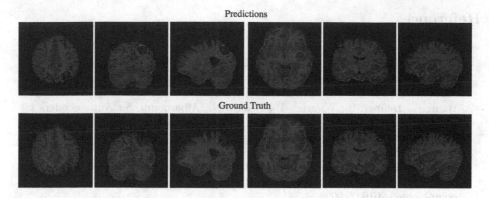

Ground Truth

Fig. 3. Predictions of our final model with AutoSeg and 2.5D training for artificial anomalies on the held-out test set of MOOD brain MRI. Left: uniform shift, right: noise addition.

prediction. Here, the shift created a dent in the outer brain surface. The model predicts the anomaly only on the brain matter, missing the dent.

Figure 2 shows the localization quality of our model for real-world anomalies on the example of cardiomegaly. Although being by far not as accurate as for artificial anomalies, the model is able to identify the heart as the source of the anomaly.

6   Discussion

Our proposed method steers the inductive bias of a model towards detecting pathologies – instead of general statistical anomalies – by generating artificial anomalies that better represent anomalies in the real world. We have shown that our method produces impressive results on unseen artificial anomalies, outperforming existing methods by a large margin. However, real-world pathologies have potentially very different properties. This can be seen in the ChestX-ray14 dataset, where – despite our model outperforming all competing methods – the results are far from being usable, even in such a benchmark task. When faced with well-known practical problems like distribution shift, the performance might easily degenerate completely. Likewise, the results in the MOOD challenge don't reflect the performance our model achieves on our artificial anomalies, indicating that samples in the hidden test set have different properties than our train anomalies.

Nevertheless, our results show that steering the inductive bias towards the actual goal of detecting pathologies improves their detection performance. These findings motivate us to search for better anomalies with characteristics closer to real-world pathologies in future work.

References

1. Atlason, H.E., Love, A., Sigurdsson, S., Gudnason, V., Ellingsen, L.M.: Unsupervised brain lesion segmentation from MRI using a convolutional autoencoder. In: Medical Imaging 2019: Image Processing, March 2019. https://doi.org/10.1117/12.2512953
2. Baur, C., Denner, S., Wiestler, B., Navab, N., Albarqouni, S.: Autoencoders for unsupervised anomaly segmentation in brain MR images: a comparative study. Med. Image Anal. **69**, 101952 (2021)
3. Buda, M., Saha, A., Mazurowski, M.A.: Association of genomic subtypes of lower-grade gliomas with shape features automatically extracted by a deep learning algorithm. Comput. Biol. Med. **109**, 218–225 (2019). https://doi.org/10.1016/j.compbiomed.2019.05.002
4. Loshchilov, I., Hutter, F.: Decoupled weight decay regularization (2019)
5. Meissen, F., Kaissis, G., Rueckert, D.: Challenging current semi-supervised anomaly segmentation methods for brain MRI (2021)
6. Paszke, A., et al.: PyTorch: an imperative style, high-performance deep learning library. In: Advances in Neural Information Processing Systems vol. 32, pp. 8024–8035. Curran Associates, Inc. (2019). http://papers.neurips.cc/paper/9015-pytorch-an-imperative-style-high-performance-deep-learning-library.pdf
7. Perslev, M., Dam, E.B., Pai, A., Igel, C.: One network to segment them all: a general, lightweight system for accurate 3D medical image segmentation. In: Shen, D., et al. (eds.) MICCAI 2019. LNCS, vol. 11765, pp. 30–38. Springer, Cham (2019). https://doi.org/10.1007/978-3-030-32245-8_4
8. Petersen, J., et al.: Medical out-of-distribution analysis challenge, March 2021. https://doi.org/10.5281/zenodo.4573948
9. Pinaya, W.H.L., et al.: Unsupervised brain anomaly detection and segmentation with transformers (2021)
10. Schlegl, T., Seeböck, P., Waldstein, S.M., Langs, G., Schmidt-Erfurth, U.: f-AnoGAN: fast unsupervised anomaly detection with generative adversarial networks. Med. Image Anal. **54**, 30–44 (2019). https://doi.org/10.1016/j.media.2019.01.010
11. Smith, K., et al.: Data from CT colonography. The cancer imaging archive (2015). https://doi.org/10.7937/K9/TCIA.2015.NWTESAY1
12. Sun, D., et al.: Autoflow: learning a better training set for optical flow. In: CVPR (2021)
13. Tan, J., Hou, B., Batten, J., Qiu, H., Kainz, B.: Detecting outliers with foreign patch interpolation (2020)
14. Van Essen, D.C., Smith, S.M., Barch, D.M., Behrens, T.E., Yacoub, E., Ugurbil, K.: The WU-Minn human connectome project: an overview. NeuroImage **80**, 62–79 (2013)
15. Wang, X., Peng, Y., Lu, L., Lu, Z., Bagheri, M., Summers, R.M.: ChestX-ray: hospital-scale chest X-ray database and benchmarks on weakly supervised classification and localization of common thorax diseases. In: Lu, L., Wang, X., Carneiro, G., Yang, L. (eds.) Deep Learning and Convolutional Neural Networks for Medical Imaging and Clinical Informatics. ACVPR, pp. 369–392. Springer, Cham (2019). https://doi.org/10.1007/978-3-030-13969-8_18

16. Wood, E., et al.: Fake it till you make it: face analysis in the wild using synthetic data alone (2021)
17. Zimmerer, D., Isensee, F., Petersen, J., Kohl, S., Maier-Hein, K.: Unsupervised anomaly localization using variational auto-encoders. In: Shen, D., et al. (eds.) MICCAI 2019. LNCS, vol. 11767, pp. 289–297. Springer, Cham (2019). https://doi.org/10.1007/978-3-030-32251-9_32

L2R

Learn2Reg 2021 Preface

Medical image registration continues to play a very important role in improving clinical workflows, computer-assisted interventions, and diagnosis as well as for research studies involving, for example, morphological analysis. Over the last few years deep learning for medical registration has significantly advanced in terms of robustness, computation speed, and accuracy and is often able to match or outperform conventional algorithms. This was demonstrated at our first Learn2Reg benchmark evaluation at MICCAI 2020. Due to their ease of implementation and fast inference speeds GPU-accelerated algorithms are likely to improve practical translation. Nevertheless, there are many more clinically useful applications that have so far not been evaluated in a comparative setting that provides an unbiased benchmark across learning-based registration methods, comparing such methods with each other and with their conventional (not trained) counterparts. This applies in particular to abdominal/thorax MRI/CT registration and whole-brain inter-subject alignment of MRI. Akin to Learn2Reg 2020, this new edition also aimed at providing standardised datasets that were easily available and accessible and resulted in a simplified challenge design that removed many of the common pitfalls for learning and applying transformations.

Our challenge comprised three clinically relevant sub-tasks (datasets) that were complementary in nature. Participants could either individually or comprehensively address these tasks that covered both intra- and inter-patient alignment; CT and MRI modalities; and neuro-, thorax, and abdominal anatomies registration and address many imminent challenges of medical image registration: learning from small datasets; estimating large deformations; dealing with multi-modal scans; and learning from limited annotations.

A total of more than 500 annotated 3D scans were made available to the public, including 32 inter-patient abdominal MRI/CTs, 30 pairs of inspiration and exhale lung CTs, and over 400 whole-brain MRI scans (for details please see learn2reg.grand-challenge.org/Datasets). The evaluation of the more than two dozen individual task submissions was carried out with a comprehensive evaluation pipeline, based on displacement fields, to compute the methods' performances. Since medical image registration is not limited to accurately and robustly transferring anatomical annotations, which was measured by computing target registration errors of landmarks or Dice and surface metrics of anatomical segmentations, we also incorporated a measure of transformation complexity (the standard deviation of local volume change defined by the log Jacobian determinant of the deformation) and a direct comparison of run times through either CPU or our provided Nvidia GPU backends (this was not a strict requirement for participants).

All metrics were converted into significant ranks and an overall winner across the three tasks was determined in the method described in "Conditional Deep Laplacian Pyramid Image Registration Network in Learn2Reg Challenge" by Tony Mok and Albert Chung. Some interesting outcomes were that combinations of feed-forward prediction and iterative instance optimization were used by multiple top-ranked approaches and that de-coupling semantic feature learning in combination with conventional optimization yielded excellent results. A detailed analysis of comparative results from both 2020 and 2021 can be found at: https://arxiv.org/pdf/2112.04489.pdf [1].

The proceedings of this workshop contain eight selected papers that cover a wide spectrum of conventional and learning-based registration methods and often describe novel contributions. All papers underwent a light review process by our Program Committee chairs. We would like to thank all the Learn2Reg participants and co-organizers for their efforts that helped provide substantial new insights for this emerging research field and immensely contributed to the success of this challenge.

January 2022

Mattias P. Heinrich
Alessa Hering
Lasse Hansen
Adrian Dalca

Reference

1. Hering, A., Hansen, L., Mok, T.C.W., Chung, A.C.S., Siebert, H., Häger, S., Lange, A., Kuckertz, S., Heldmann, S., Shao, W., Vesal, S., Rusu, M., Sonn, G., Estienne, T., Vakalopoulou, M., Han, L., Huang, Y., Brudfors, M., Balbastre, Y., Joutard, S., Modat, M., Lifshitz, G., Raviv, D., Lv, J., Li, Q., Jaouen, V., Visvikis, D., Fourcade, C., Rubeaux, M., Pan, W., Xu, Z., Jian, B., Benetti, F.D., Wodzinski, M., Gunnarsson, N., Sjölund, J., Qiu, H., Li, Z., Großbröhmer, C., Hoopes, A., Reinertsen, I., Xiao, Y., Landman, B., Huo, Y., Murphy, K., Lessmann, N., van Ginneken, B., Dalca, A.V., Heinrich, M.P.: Learn2reg: comprehensive multi-task medical image registration challenge, dataset and evaluation in the era of deep learning (2021)

Learn2Reg 2021 Organization

General Chair

Mattias Heinrich   University of Lübeck, Germany

Program Committee Chairs

Alessa Hering   Radboud UMC, The Netherlands,
                                       and Fraunhofer MEVIS, Germany
Lasse Hansen   University of Lübeck, Germany

Advisor

Adrian Dalca   A.A. Martinos Center for Biomedical
                                       Imaging MGH and MIT, USA

Data and Technical Contributors

Bram van Ginneken   Radboud UMC, The Netherlands
Adrian Dalca   A.A. Martinos Center for Biomedical
                                       Imaging MGH and MIT, USA
Christoph Großbröhmer   University of Lübeck, Germany
Hanna Sieber   University of Lübeck, Germany
Lasse Hansen   University of Lübeck, Germany

Deformable Registration of Brain MR Images via a Hybrid Loss

Luyi Han[1], Haoran Dou[2], Yunzhi Huang[3(✉)], and Pew-Thian Yap[4(✉)]

[1] Department of Radiology and Nuclear Medicine, Radboud University Medical Center, Geert Grooteplein 10, 6525 Nijmegen, GA, The Netherlands
[2] Centre for Computational Imaging and Simulation Technologies in Biomedicine (CISTIB), University of Leeds, Leeds, UK
[3] College of Biomedical Engineering, Sichuan University, Chengdu, China
[4] Department of Radiology and Biomedical Research Imaging Center (BRIC), University of North Carolina, Chapel Hill, USA
ptyap@med.unc.edu

Abstract. Unsupervised learning strategy is widely adopted by the deformable registration models due to the lack of ground truth of deformation fields. These models typically depend on the intensity-based similarity loss to obtain the learning convergence. Despite the success, such dependence is insufficient. For the deformable registration of mono-modality image, well-aligned two images not only have indistinguishable intensity differences, but also are close in the statistical distribution and the boundary areas. Considering that well-designed loss functions can facilitate a learning model into a desirable convergence, we learn a deformable registration model for T1-weighted MR images by integrating multiple image characteristics via a hybrid loss. Our method registers the OASIS dataset with high accuracy while preserving deformation smoothness.

1 Introduction

Deformable registration estimates dense deformation fields to establish image-to-image correspondence. Conventional methods typically involve time-consuming iterative optimization and experience-dependent parameter tuning. Alternatively, deformations can be learned for fast registration via (1) supervised learning [3,4,10,11]; (2) weakly-supervised learning [8]; and (3) unsupervised learning [2,7].

Supervised learning methods rely on deformations predicted using conventional methods (*e.g.*, SyN [1] or Diffeomorphic Demons [12]) and simulations [3,4]. In contrast, weakly-supervised and unsupervised learning methods do not require ground truth deformations. Weakly-supervised learning methods optimize model parameters via supervision using label-level similarity and segmentation maps to align structural boundaries [8]. Unsupervised learning methods are supervised via intensity-level similarity (*e.g.*, Normalized Cross Correlation (NCC) or Sum of Squared Difference (SSD)).

© Springer Nature Switzerland AG 2022
M. Aubreville et al. (Eds.): MIDOG 2021/MOOD 2021/L2R 2021, LNCS 13166, pp. 141–146, 2022.
https://doi.org/10.1007/978-3-030-97281-3_20

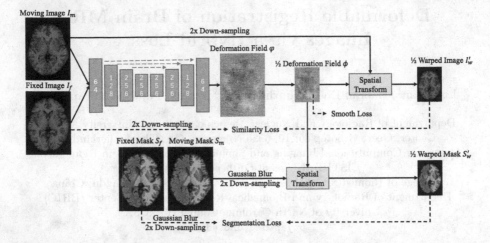

Fig. 1. Overview of the proposed registration model. Random selected patches from the inter-subject T1w image pairs input the registration network to output the deformation field. During training, similarity loss, segmentation loss, and smooth loss are bonded together to guide the learning.

Our method combines weakly-supervised and unsupervised learning and learns registration via multiple aspects, including intensity, statistics, label levels. The proposed method ranked fifth on the brain T1w deformable registration task organized by the MICCAI 2021 Learn2Reg challenge[1].

2   Method

The core of our deformable registration model (Fig. 1) is based on VoxelMorph [2] with the following modifications: (1) increased feature channels for each layer and (2) deformation field downsampling by a factor of 2. The input is a randomly selected pair of patches from the moving and fixed images. The output is the predicted x, y, and z displacements at half the resolution of the input.

2.1   Hybrid Loss

Aligned images should be matched at the boundary, intensity, and statistical distribution levels. We employ multi-faceted supervision involving a hybrid loss function to improve the alignment between the moving image I_m and the fixed image I_f.

Intensity Loss We employ the commonly used SSD to gauge the intensity dissimilarity between I_m and I_f:

$$\mathcal{L}_i = ||I_w' - I_f'||_2^2 \tag{1}$$

[1] https://learn2reg.grand-challenge.org.

where $I_w^{'} = I_m^{'} \circ \phi$ is the half size moving image $I_m^{'}$ warped with predicted displacement field ϕ. $I_m^{'}$ and $I_f^{'}$ are downsampled from the original moving image I_m and fixed image I_f by factor 2.

Statistic Loss We employ mutual information [5] to improve the joint probability distribution between $I_w^{'}$ and $I_f^{'}$:

$$\mathcal{L}_s = H(I_w^{'}) + H(I_f^{'}) - H(I_w, I_f^{'}) \tag{2}$$

where $H(\cdot)$ refers to the entropy of an image, and $H(\cdot, \cdot)$ is the joint entropy of two images.

Boundary Loss. We employ the area overlap between the segmentation mask $S_w^{'}$ of $I_w^{'}$ and downsampled segmentation mask $S_f^{'}$ of $I_f^{'}$ for boundary-level supervision. The segmentation maps are encoded in one-hot format and are convoluted with a Gaussian blur kernel with the size of 7 and σ of 1. We combine $L1$ and *Dice* for boundary loss:

$$\mathcal{L}_b = \|S_w^{'} - S_f^{'}\|_1 + (1 - \frac{2\|S_w^{'} \cdot S_f^{'}\|_1}{\|S_w^{'} + S_f^{'}\|_1}) \tag{3}$$

where $S_w^{'} = S_m^{'} \circ \phi$ refers to the warped segmentation map.

Total Loss. In addition to the losses described above, we include a gradient-based regularization term to preserve the topology of the deformation field. The total loss is

$$\mathcal{L} = \mathcal{L}_i + \mathcal{L}_s + \mathcal{L}_b + \lambda \cdot \mathtt{Grad}(\phi) \tag{4}$$

where λ balances the dissimilarity term and regularization term and is set empirically to 0.8 via grid research.

2.2   Dataset and Implementation Details

Dataset Training (414 subjects), validation (20 subjects), and testing (39 subjects) were based on the Open access series of imaging studies (OASIS) dataset [9] curated by the organizers of Learn2Reg MICCAI Challenge 2021 [6][2]. OASIS is a cross-sectional MRI data study with a wide range of participants from young, middle aged, nondemented, and demented older adults. Pre-processing (skull-stripping, normalisation, pre-alignment and resampling) was done according to the procedure described in [7]. Semi-automatic labels with manual corrections of 35 cortical and subcortical brain structures were generated using 3D Slicer.

---

[2] https://learn2reg.grand-challenge.org/Datasets/.

Implementation Details. We implemented our method using Pytorch with NIVIDIA 3090 RTX. We optimized the model with ADAM, with learning rate $1e-6$, a default of 200,000 steps, and batch size of 1. During training, data was augmented by randomly selecting patches of size $128 \times 128 \times 128$ from the input volumes. During testing, a deformation field was predicted for an image volume. 13GB and 11GB of GPU memory was consumed during the training and testing stages, respectively.

3   Experimental Results

3.1   Results

Table 1 lists the registration accuracy of different settings on the loss functions. Our method achieves for the testing dataset an average Dice score of 80.47% with standard error 1.67% and an average Hausdorff distance of $1.8015 \pm 0.4325mm$ over 35 brain ROIs, with SDlogJ of 0.0822 ± 0.0042 for full size deformation field ψ. Figure 2 shows exemplar registration results given by our method.

Table 1. Ablation study on the validation dataset.

Method	DSC↑	HD(mm)↓	SDlogJ↓
patch VM-c32	0.7978±0.0230	1.9733±0.4777	0.0848±0.0057
patch VM-c64	0.8040±0.0209	1.9432±0.4687	0.0839±0.0053
patch VM-c64+MI+Dice	0.8117±0.0214	1.8549±0.4363	0.0811±0.0053
patch VM-c64+MI+Dice+halved(Proposed)	0.8395±0.0142	1.6635±0.3734	0.0788±0.0044

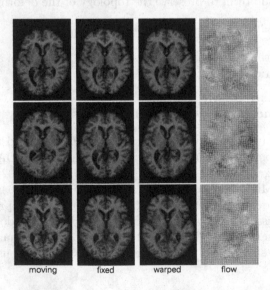

moving   fixed   warped   flow

Fig. 2. From left to right are the moving image, the fixed image, the warped image, and the predicted deformation field.

4 Discussion and Conclusion

A good deformable registration framework is required to accurately calculate the complex mapping between image pairs. The widely used intensity similarity loss is highly dependent on the image quality, requiring the voxels inside each tissue to close at intensity across individual data. Hence, the intensity-driven supervision is efficient yet insufficient. A diverse similarity constraints is then required to enhance the optimization at multiple levels.

In this Challenge, we adopted a hybrid similarity loss to steer the learning procedure. The intensity-based SSD loss and the statistic-based MI loss steer the registration accuracy at both the local voxels and the global intensity distributions. Meanwhile, the boundary loss improves the match at the boundary regions. We showed that registration of T1-weighted images can be registered with high accuracy by enforcing similarity at the intensity, statistic, and boundary levels.

Acknowledgment. Thanks all the organizers of the MICCAI 2021 Learn2Reg challenge. The work was supported in part by the National Natural Science Foundation of China under Grant 6210011424.

References

1. Avants, B., Epstein, C., Grossman, M., Gee, J.: Symmetric diffeomorphic image registration with cross-correlation: evaluating automated labeling of elderly and neurodegenerative brain. Med. Image Anal. **12**(1), 26–41 (2008)
2. Balakrishnan, G., Zhao, A., Sabuncu, M.R., Guttag, J., Dalca, A.V.: An unsupervised learning model for deformable medical image registration. In: Proceedings of the IEEE Conference on Computer Vision and Pattern Recognition, pp. 9252–9260 (2018)
3. Cao, X., Yang, J., Zhang, J., Wang, Q., Yap, P.T., Shen, D.: Deformable image registration using a cue-aware deep regression network. IEEE Trans. Biomed. Eng. **65**(9), 1900–1911 (2018)
4. Eppenhof, K.A., Pluim, J.P.: Pulmonary CT registration through supervised learning with convolutional neural networks. IEEE Trans. Med. Imaging **38**(5), 1097–1105 (2018)
5. Guo, C.K.: Multi-modal image registration with unsupervised deep learning. Ph.D. thesis, Massachusetts Institute of Technology (2019)
6. Hering, A., et al.: Learn2reg: comprehensive multi-task medical image registration challenge, dataset and evaluation in the era of deep learning (2021)
7. Hoopes, A., Hoffmann, M., Fischl, B., Guttag, J., Dalca, A.V.: HyperMorph: amortized hyperparameter learning for image registration. In: Feragen, A., Sommer, S., Schnabel, J., Nielsen, M. (eds.) IPMI 2021. LNCS, vol. 12729, pp. 3–17. Springer, Cham (2021). https://doi.org/10.1007/978-3-030-78191-0_1
8. Hu, Y., et al.: Weakly-supervised convolutional neural networks for multimodal image registration. Med. Image Anal. **49**, 1–13 (2018)
9. Marcus, D.S., Wang, T.H., Parker, J., Csernansky, J.G., Morris, J.C., Buckner, R.L.: Open access series of imaging studies (oasis): cross-sectional MRI data in young, middle aged, nondemented, and demented older adults. J. Cogn. Neurosci. **19**(9), 1498–1507 (2007)

10. Rohé, M.-M., Datar, M., Heimann, T., Sermesant, M., Pennec, X.: SVF-Net: learning deformable image registration using shape matching. In: Descoteaux, M., Maier-Hein, L., Franz, A., Jannin, P., Collins, D.L., Duchesne, S. (eds.) MICCAI 2017. LNCS, vol. 10433, pp. 266–274. Springer, Cham (2017). https://doi.org/10.1007/978-3-319-66182-7_31
11. Sokooti, H., de Vos, B., Berendsen, F., Lelieveldt, B.P.F., Išgum, I., Staring, M.: Nonrigid image registration using multi-scale 3D convolutional neural networks. In: Descoteaux, M., Maier-Hein, L., Franz, A., Jannin, P., Collins, D.L., Duchesne, S. (eds.) MICCAI 2017. LNCS, vol. 10433, pp. 232–239. Springer, Cham (2017). https://doi.org/10.1007/978-3-319-66182-7_27
12. Vercauteren, T., Pennec, X., Perchant, A., Ayache, N.: Diffeomorphic demons: efficient non-parametric image registration. NeuroImage 45(1, Supplement 1), S61–S72 (2009)

Fraunhofer MEVIS Image Registration Solutions for the Learn2Reg 2021 Challenge

Alessa Hering$^{(\boxtimes)}$, Annkristin Lange , Stefan Heldmann ,
Stephanie Häger , and Sven Kuckertz

Fraunhofer Institute for Digital Medicine MEVIS, Lübeck, Germany
alessa.hering@mevis.fraunhofer.de

Abstract. In this paper, we present our contribution to the learn2reg challenge. We applied the Fraunhofer MEVIS registration library RegLib comprehensively to all 3 tasks of the challenge, where we used a classic iterative registration method with NGF distance measure, second order curvature regularizer and a multi-level optimization scheme. We show that with our proposed method robust results can be achieved throughout all tasks resulting in the fourth place overall task and the best accuracy on the lung CT registration task.

Keywords: Image Registration · Registration Challenge · Learn2Reg

1 Introduction

Image registration is a key task in medical image analysis to estimate deformations between images and to obtain spatial correspondences. The goal of image registration is to find a reasonable deformation for a pair of fixed and moving image so that the transformed moving image and the fixed image are similar. Image registration is typically formulated as an optimization problem where a suitable cost function is minimized through iterative optimization schemes. Over time, a variety of image registration models and approaches have been developed. Therefore comparison possibilities are needed. In order to ensure comparability, challenges are created in which the different registration procedures are evaluated on the same image data and under the same computation conditions. One such challenge is the *Learn2Reg: 2021 MICCAI Registration Challenge* [7]. It consists of 3 different registration tasks that cover both intra- and inter-patient alignment, CT and MRI modalities, neuro-, thorax and abdominal anatomies. In this paper we present our solutions to all 3 tasks of the challenge.

2 Method and Results

All 3 tasks are solved by classical iterative methods and build on cost functions and losses made up from several terms that are selected for the specific task.

© Springer Nature Switzerland AG 2022
M. Aubreville et al. (Eds.): MIDOG 2021/MOOD 2021/L2R 2021, LNCS 13166, pp. 147–152, 2022.
https://doi.org/10.1007/978-3-030-97281-3_21

Common to all is the use of normalized gradient fields (NGF) [5] image similarity for fixed and moving images $\mathcal{F}, \mathcal{M} : \Omega \subset \mathbb{R}^3 \to \mathbb{R}$

$$\text{NGF}(\mathcal{F}, \mathcal{M}) = \frac{1}{2} \int_\Omega 1 - \frac{\langle \nabla \mathcal{F}, \nabla \mathcal{M} \rangle^2_{\epsilon_\mathcal{F} \epsilon_\mathcal{M}}}{\|\nabla \mathcal{F}\|^2_{\epsilon_\mathcal{F}} \|\nabla \mathcal{M}\|^2_{\epsilon_\mathcal{M}}} \, dx \qquad (1)$$

with parameters $\epsilon_\mathcal{F}, \epsilon_\mathcal{M} > 0$, $\langle x, y \rangle_\epsilon := x^\top y + \epsilon$ and $\|x\|_\epsilon = \sqrt{\langle x, y \rangle_\epsilon}$ and 2nd order curvature (CURV) regularization [4] of displacement vector fields $u : \Omega \subset \mathbb{R}^3 \to \mathbb{R}^3$

$$\text{CURV}(u) = \frac{1}{2} \int_\Omega \sum_{\ell=1}^3 \|\Delta u_\ell\|^2 \, dx. \qquad (2)$$

Furthermore, the methods use a coarse-to-fine multi-level iterative registration scheme where a Gaussian image pyramid is generated for both images to obtain downsampled and smoothed images. Then, a registration is performed on the lowest resolution level and the resulting deformation field serves as the starting point for the following registration on the next highest level. This proceeds till the finest level with quasi-Newton L-BFGS optimization at each level. This procedure allows to align larger structures on the lower levels and helps to avoid local minima, to reduce topological changes or foldings, and to speed up run times.

Metrics for accuracy (TRE, DICE, Hausdorff95), robustness (DICE30, TRE30, DICEunknownLabel) and plausibility of the deformation field (Log-JacDetStd) are computed for evaluation of the challenge as well as the runtimes. More details can be found in [1]. Table 1 shows the results of our methods for all 3 tasks.

Task 1

The aim of the first task was the intra-patient registration of abdominal MR and CT scans [3]. 8 training and 8 test MR-CT pairs were provided with preprocessing such as same isotropic voxel resolutions (2 mm) and spatial dimensions ($192 \times 160 \times 192$) as well as affine preregistration. In addition, areas of interest (ROI) were included for both MR and CT scans, as well as for training and test images. Although the images are affine preregistered, we performed a rigid registration of the organ ROIs as a first step to obtain a better starting point for subsequent nonparametric registration. Since the ROIs are masks, the SSD distance measure was chosen. The deformable registration is performed twice in our approach, first with $\alpha = 50$ to give more weight to the regularization. Then, the same registration is performed again, using the deformation field of the first registration as initial value, this time with $\alpha = 15$ to give a better fit to the details. As edge parameter ϵ we have chosen half of the average image gradient in each case. For both deformable registrations we used a multi-level optimization scheme with 3 levels. As shown in Table 1 the DICE score could be improved from 0.23 to 0.71 on the test CT scans. One difficulty was that in 3 test cases an organ was missing in the CT images. The ROIs for MRI and CT are therefore different and the rigid registration of the ROIs did not work properly

Table 1. Results of our methods in the Learn2Reg Challenge. For the challenge the target registration error (TRE), DICE score, 95% percentile of the Hausdorff distance (Hausdorff95) and the standard deviation of log Jacobian determinant of the deformation field (LogJacDetStd) were measured. We additionally measured the runtimes for our methods on a machine of the challenge organizers using a docker container.

	task 1	task 2	task3
TRE before reg. [mm]	-	10.24 ± 5.28	-
TRE after reg. [mm]	-	1.68 ± 2.31	-
TRE30 before reg. [mm]	-	16.80	-
TRE30 after reg. [mm]	-	2.37	-
DICE before reg	0.23 ± 0.19	-	0.56 ± 0.21
DICE after reg	0.71 ± 0.16	-	0.77 ± 0.17
DICE unknown label after reg	0.65	-	0.59
DICE30 before reg	0.23	-	0.56
DICE30 after reg	0.67	-	0.77
Hausdorff95 before reg. [mm]	42.18 ± 13.55	-	3.85 ± 1.89
Hausdorff95 after reg. [mm]	21.04 ± 14.07	-	2.08 ± 1.7
LogJacDetStd	0.15 ± 0.04	0.08 ± 0.07	0.07 ± 0.01
runtime [s]	14.73 ± 1.46	95.38 ± 18.01	10.36 ± 0.52

and led to unwanted rotations. In Fig. 1 an exemplary registration result on the validation data with the overlayed segmentations is shown.

Task 2
The aim of the second task was the registration of expiration to inspiration CT scans of the lung. The provided data consists of 20 training scan pairs [9] and 10 test scan pairs [8]. All scan pairs were resampled to a image size of $192 \times 192 \times 208$ and were affine pre-registered. The main challenges are the large deformation due to breathing and that the lungs in the expiration scans are not fully visible.

Our submitted method based on our previous work [12]. First, a graph-based matching of a large number of keypoints for the estimation of robust large-motion correspondences is performed. Then, this is followed by a continuous, deformable image registration incorporating both image intensities and keypoint information. Herefore, we used the NGF distance measure with edge parameter $\epsilon = 0.1$. For a smooth deformation field we selected the curvature regularizer with weight parameter $\alpha = 1$. In contrast to [12], we are not integrating the lung mask into a cost term to enforce lung boundary alignment, because the expiration lung is not fully visible. However, we mask the NGF distance measure with the expiration lung mask. A coarse-to-fine multilevel scheme with 3 levels was applied. In contrast to last year's submission [6], we increased the grid from $33 \times 33 \times 33$ to $55 \times 55 \times 55$. With this change, the target registration error decreases but also the regularity of the deformation field. With a target

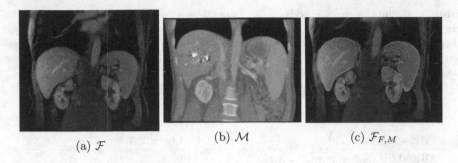

(a) $\mathcal{F}$ (b) $\mathcal{M}$ (c) $\mathcal{F}_{F,M}$

Fig. 1. Exemplary registration result for task 1. The original fixed (MR) and moving (CT) images are shown in coronal direction (a, b). In (c) the fixed MR is overlayed by the fixed (green) and deformed moving (orange) labels. (Color figure online)

(a) $\mathcal{F}$ (b) $\mathcal{M}$ (c) $|\mathcal{F} - \mathcal{M}|$ (d) $|\mathcal{F} - \mathcal{M}(y)|$

Fig. 2. Example coronal slices extracted from a exemplary case for task 2: a) The expiration image, b) inspiration image, c) the difference image before the registration and d) the difference image after registration. For a better visualization, we only show the image inside the lung, however, the full thorax scan was used.

registration error of 1.6 ± 2.311 mm, we archived the highest accuracy of all submissions in the challenge. The whole registration pipeline takes about 92.7 s which includes the keypoint detection with 86.4 s and the actual registration with 6 s. All results are summarized in Table 1. To illustrate the registration results, we show the difference images $\mathcal{F} - \mathcal{M}(y)$ before and after registration in Fig. 2. The breathing motion was successfully recovered and inner lung structures are well aligned.

Task 3

The third task deals with the challenge of inter-patient registration of whole brain MR scans. The data for this task is provided preprocessed including affine prealignment and resampling, resulting in images with a size of $160 \times 192 \times 224$ voxels at an isotropic 1mm resolution. Additionally, segmentation masks for 35 brain structures are available. The alignment of these predominantly small structures of variable shape and size form a challenging task, especially between different patients. The dataset consists of 414 training images including images for validation [10,11].

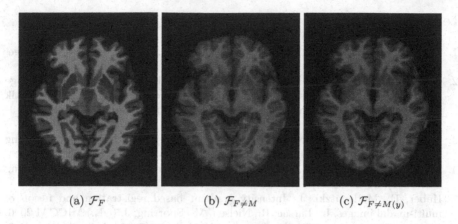

(a) $\mathcal{F}_F$ (b) $\mathcal{F}_{F \neq M}$ (c) $\mathcal{F}_{F \neq M(y)}$

Fig. 3. Exemplary registration result for task 3. The fixed MRI and its corresponding labels are shown in axial direction (a). The fixed image is overlayed by the difference of fixed and moving labels before (b) and after the registration (c), respectively.

In order to speed up the computation we use a fast implementation of our baseline algorithm on the GPU [2]. This results in a time for optimization of under a second and a total runtime of around 10 s including data loading and postprocessing. For optimization of the registration problem we use the NGF distance measure with parameters $\epsilon_F = 0.01$ and $\epsilon_M = 0.01$ which are additionally multiplied by the average gradient for more robustness. The multi-level pyramid consists of three stages, where the output has half the resolution of the input images.

As shown in Table 1, our method is able to improve the DICE score on average from 0.56 to 0.77, while maintaining physically plausible results with a log Jacobian determinant of 0.07. Moreover, the 30% lowest DICE score of all cases is the same as the average score (0.77), which shows that our algorithm is robust against outliers. An exemplary registration result for this task is shown in Fig. 3.

3 Conclusion

We showed that the Fraunhofer MEVIS RegLib is successfully applicable to all three tasks of the Learn2Reg challenge that differ greatly and cover both intra- and inter-patient alignment, various modalities and anatomies. We chose the classic iterative method for all 3 tasks and achieved the forth place in the challenge. Furthermore, we achieved the overall highest registration accuracy with our method in task 2.

References

1. Learn2Reg challenge, metrics and evaluation. https://learn2reg.grand-challenge.org/Submission/
2. Budelmann, D., König, L., Papenberg, N., Lellmann, J.: Fully-deformable 3D image registration in two seconds. In: Bildverarbeitung für die Medizin 2019. Informatik aktuell, pp. 302–307. Springer, Wiesbaden (2019). https://doi.org/10.1007/978-3-658-25326-4_67
3. Clark, K., et al.: The cancer imaging archive (TCIA): maintaining and operating a public information repository. J. Digit. Imaging **26**(6), 1045–1057 (2013)
4. Fischer, B., Modersitzki, J.: Curvature based image registration. J. Math. Imaging Vis. **18**(1), 81–85 (2003)
5. Haber, E., Modersitzki, J.: Intensity gradient based registration and fusion of multi-modal images. In: Larsen, R., Nielsen, M., Sporring, J. (eds.) MICCAI 2006. LNCS, vol. 4191, pp. 726–733. Springer, Heidelberg (2006). https://doi.org/10.1007/11866763_89
6. Häger, S., Heldmann, S., Hering, A., Kuckertz, S., Lange, A.: Variable fraunhofer MEVIS RegLib comprehensively applied to Learn2Reg challenge. In: Shusharina, N., Heinrich, M.P., Huang, R. (eds.) MICCAI 2020. LNCS, vol. 12587, pp. 74–79. Springer, Cham (2021). https://doi.org/10.1007/978-3-030-71827-5_9
7. Hering, A., et al.: Learn2Reg: comprehensive multi-task medical image registration challenge, dataset and evaluation in the era of deep learning. arXiv preprint arXiv:2112.04489 (2021)
8. Hering, A., Murphy, K., van Ginneken, B.: Learn2Reg challenge: CT lung registration - test data, September 2020. https://doi.org/10.5281/zenodo.4048761
9. Hering, A., Murphy, K., van Ginneken, B.: Learn2Reg challenge: CT lung registration - training data, May 2020. https://doi.org/10.5281/zenodo.3835682
10. Hoopes, A., Hoffmann, M., Fischl, B., Guttag, J.V., Dalca, A.V.: Hypermorph: Amortized hyperparameter learning for image registration. CoRR abs/2101.01035 (2021). https://arxiv.org/abs/2101.01035
11. Marcus, D.S., et al.: Open access series of imaging studies (OASIS): cross-sectional MRI data in young, middle aged, nondemented, and demented older adults. J. Cognit. Neurosci. **19**(9), 1498–1507 (2007)
12. Rühaak, J., et al.: Estimation of large motion in lung CT by integrating regularized keypoint correspondences into dense deformable registration. IEEE Trans. Med. Imaging **36**(8), 1746–1757 (2017)

Unsupervised Volumetric Displacement Fields Using Cost Function Unrolling

Gal Lifshitz$^{(\boxtimes)}$ and Dan Raviv

Tel Aviv University, Tel Aviv, Israel
lifshitz@mail.tau.ac.il, darav@tauex.tau.ac.il

Abstract. Steepest descent algorithms, which are commonly used in deep learning, use the gradient as the descent direction, either as-is or after a direction shift using preconditioning. In many scenarios calculating the gradient is numerically hard due to complex or non-differentiable cost functions, specifically next to singular points. In this work, we focus on the derivation of the Total Variation regularizer commonly used in unsupervised displacement fields cost functions. Specifically, we derive a differentiable proxy to the hard L^1 smoothness constraint in an iterative scheme, which we refer to as Cost Unrolling. We show that our unrolled cost function enables more accurate gradients in regions where the gradients are hard to evaluate or even undefined without increasing the complexity of the original model. We demonstrate the effectiveness of our method in synthetic tests, as well as in the task of unsupervised learning of displacement fields between corresponding 3DCT lung scans. We report improved results compared to standard TV in all tested scenarios, achieved without modifying model architecture but simply through improving the gradients during training.

Keywords: Unsupervised Learning · Displacement Fields · Optimization

1 Introduction

The L^1 norm of the gradients of a given function, also known as Total Variation (TV), and more specifically its estimation, has been a significant field of study in robust statistics [11]. Even prior to the sweeping AI era, many approaches to Computer Vision problems, such as image restoration, denoising and registration [18,20] have used a TV regularizer, as it represents the prior distribution of pixel intensities of natural images [10]. Its main advantage is its robustness to small oscillations such as noise while preserving sharp discontinuities such as edges.

Historically, solving the TV problem has been a challenging task, mainly due to the non-differentiability of the L^1 norm at zero. Early approaches consisted of differentiable approximations [4], however iterative variational methods [2,18,20] have been shown superior.

Introducing trainable Deep Neural Networks (DNNs) to tackle Computer Vision tasks has brought a significant performance boost, and specifically, the

© Springer Nature Switzerland AG 2022
M. Aubreville et al. (Eds.): MIDOG 2021/MOOD 2021/L2R 2021, LNCS 13166, pp. 153–160, 2022.
https://doi.org/10.1007/978-3-030-97281-3_22

commonly used auto-derivation frameworks [1,17] have provided us with quick and easy tools to solve complex functions. Indeed, these frameworks may bypass the L^1 non-differentiability either by a differentiable proxy or using its sub-gradients. Not surprisingly, the TV smoothness regularization can be found in the cost functions of many 2D and 3D image registration works both past and recent [9,19]. We claim that non-differentiable cost functions should be dealt with greater care, as was done for many years before the deep learning era.

Cost Unrolling, a novel pipeline, in which the commonly used unsupervised cost function consisting of a data term and TV smoothness regularization is unrolled to obtain an iterative proxy, has been introduced in [13]. Following the well-known Alternating Direction Method of Multiplies (ADMM) [2] algorithm, the hard initial optimization problem is iteratively decomposed into a set of sub-problems, each one featuring a differentiable cost function. Gradients accumulated using all sub-cost functions at each training step have been shown to be more accurate in the regions where the gradients of the original cost function are hard to evaluate or undefined, improving convergence. In this paper, we expand Cost Unrolling to the dense 3D domain. Specifically, testing our method on unsupervised image registration tasks, using both synthetic data and real-world raw lung 3DCT scans, we find training a DNN model using the unrolled cost improves results, convergences faster and enables smooth, yet edge-preserving displacement fields, without modifying the model architecture.

We demonstrate here that unlike all other methods, improving the ability of a model to predict more accurate displacement fields can be achieved simply through improving the computed gradients during training.

2   Cost Function Unrolling

A general formulation of the cost function used for unsupervised learning consists of a data term, measuring the likelihood of a given prediction over the given data, as well as a prior term, constraining probable predictions. In this work we consider the TV regularized unsupervised cost function case and unroll it to obtain our novel smoothness regularizer. Denote by Θ the set of trainable parameters of a DNN model, its predicted output $\mathbf{F}$ and the set of unlabeled training data $\mathcal{I}$. The unsupervised TV regularized cost function takes the form:

$$\mathcal{L}(\mathcal{I}, \Theta) = \mathbf{\Phi}(\mathbf{F}, \mathcal{I}) + \lambda \|\nabla \mathbf{F}\|_1 \qquad (1)$$

where $\mathbf{\Phi}(\cdot)$ is a differentiable function measuring the likelihood of $\mathbf{F}$ over the training data $\mathcal{I}$, $\nabla \mathbf{F}$ are its spatial gradients, $\| \cdot \|_1$ is the L^1 norm and λ is a hyperparameter controlling regularization. Note that the used L^1 norm function is non-differentiable, specifically around its optimum, and therefore we wish to obtain an iterative differentiable proxy.

2.1   Unrolling the Unsupervised Cost Function

Our goal is to minimize the objective function in (1). Following the ADMM [2] algorithm, we derive the iterative update steps minimizing (1), which are then used to construct our unrolled cost function. Introducing an auxiliary variable $\mathbf{Q}$, the Lagrange Multipliers matrix β and a penalty parameter ρ, the iterative ADMM update steps solving (1) are:

$$\hat{\mathbf{F}} = \min_{\mathbf{F}} \left\{ \Phi\left(\mathbf{F}, \mathcal{I}\right) + \frac{\rho}{2} \|\mathbf{Q} - \nabla\mathbf{F} + \beta\|_2^2 \right\} \tag{2a}$$

$$\hat{\mathbf{Q}} = \min_{\mathbf{Q}} \left\{ \lambda\|\mathbf{Q}\|_1 + \frac{\rho}{2} \|\mathbf{Q} - \nabla\mathbf{F} + \beta\|_2^2 \right\} \tag{2b}$$

$$\beta \leftarrow \beta + \eta(\mathbf{Q} - \nabla\mathbf{F}) \tag{2c}$$

with η an update rate.

2.2 Solving the Sub-optimization Problems

The solution to (2b) is given by the well known Soft Thresholding operator $\hat{\mathbf{Q}} = \mathcal{S}_{\lambda/\rho}(\nabla\mathbf{F} - \beta)$, defined:

$$\mathcal{S}_{\lambda/\rho}(x) = \begin{cases} 0, & |x| < \lambda/\rho \\ x - \frac{\lambda}{\rho}\text{sign}(x), & |x| \geq \lambda/\rho \end{cases} \tag{3}$$

The Soft Thresholding operator performs shrinkage of the input signal, thus it promotes sparse solutions. In contrast, deriving a closed form solution for the problem in (2a) is not trivial, as Φ can be any function. However, note that (2a) consists of the same data term as in (1) with the TV smoothness regularizer replaced by a softer, differentiable constraint. Recall that minimizing the TV of a function promotes sparse output gradients. In fact, (2a) yields a differentiable alternative for TV minimization, as it suggests minimizing the L^2 distance between the true and sparsified output gradients in each ADMM iteration. This realization stands in the core of our approach.

2.3 Unrolled Cost Function

Our proposed unrolled cost function takes the form:

$$\mathcal{L}^{\mathbf{F}}(\mathcal{I}, \Theta) = \Phi\left(\mathbf{F}, \mathcal{I}\right) + \frac{\rho}{2}\frac{1}{T}\sum_{t=0}^{T-1} \|\mathbf{Q}^{(t)} + \beta^{(t)} - \nabla\mathbf{F}\|_2^2 \tag{4}$$

where T is a hyperparameter stating the number of update steps carried. Being fully differentiable, specifically around its optimum, our unrolled smoothness constraint produces more accurate gradients, converges more efficiently and improves performance without increasing model complexity, as is shown in the experimentation section.

3 Experimentation

Our experimentation consists of unsupervised 2D optical flow tests, as well as the problem of unsupervised volumetric displacement fields. The availability of ground truth for our optical flow tests enables the demonstration of the effectiveness of our method.

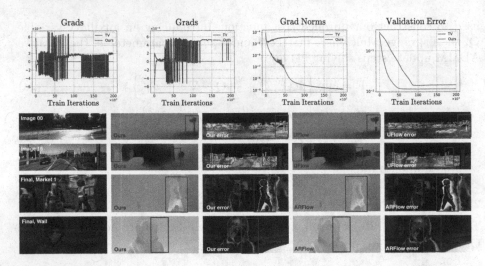

Fig. 1. Unsupervised optical flow. (top) We compare qualitative flow benchmark results of ours, the ARFlow [14] and UFlow [12] baselines. (bottom) Validation error, as well as gradients recorded during training.

3.1 2D Unsupervised Optical Flow

Tests of our unrolled cost on the unsupervised optical flow problem are detailed in [13]. Cost Unrolling is introduced to the recently published ARFlow [14] baseline, rigorously following their proposed training scheme and model architecture, yet replacing their used TV regularization with our unrolled cost. Our method is evaluated the on well-known optical flow benchmarks: the synthetic MPI Sintel [3] and autonomous driving KITTI 2015 [15]. Qualitative examples are taken from [13] and are given in Fig. 1 (bottom). Improved results are reported in all scenarios compared to standard TV. We also compare performance on the occluded regions, i.e. pixels with no correspondence, as they are highly affected by the smoothness constraint. Producing more accurate gradients during training, our method decreases the AEPE measured in occluded regions by up to 15.82% in all scenarios, enabling the detection of sharper motion edges.

Furthermore, inspecting the gradients (see Fig. 1 top) generated during training reveals that gradients of the TV suffer from severe oscillations as a result of its non-differentiability at zero. These oscillations cause the measured gradient norms to increase rather than decrease, suggesting an unstable optimum. In contrast, the gradients of our unrolled cost decrease smoothly to zero thanks to its differentiability around the optimum, enabling faster and more stable convergence.

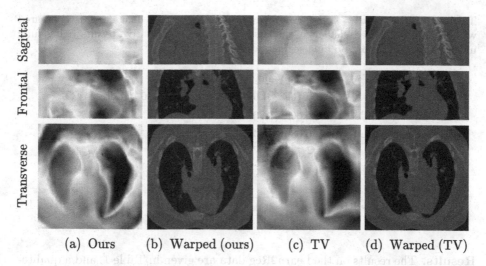

Frontal Sagittal Transverse

(a) Ours   (b) Warped (ours)   (c) TV   (d) Warped (TV)

Fig. 2. Lung 3DCT displacement fields qualitative example. Displayed are predicted displacement fields, as well as reference and warped target scans one on top of the other (color represents misaligned regions), generated using both our unrolled cost and standard TV regularization. Our method enables the capturing of motion boundaries as opposed to baseline.

3.2 Lung 3DCT Volumetric Displacement Fields

We evaluate our method on the real-world lung expiration-inspiration 3DCT unsupervised registration task, as part of the Learn2Reg challenge [8]. We expand our 2D optical flow model to the 3D domain using 3D convolution layers.

Training. Our training scheme consists of pretraining on raw Cardiac 4DCT scans, followed by finetuning on the provided training data. We train our model in a fully unsupervised manner setting Φ as the NCC between corresponding scans, using no lung segmentation masks. We used masked TV regularization replacing $\nabla \mathbf{F}$ with $\mathbf{W} \odot \nabla \mathbf{F}$ in Eq. (4), where $\mathbf{W}$ is a deterministic importance matrix, aiming to increase the penalty on the boundaries, defined:

$$\mathbf{W} = \exp\left\{-\alpha |\nabla I_1|\right\} = \exp\left\{-\alpha \left[\left|\frac{\partial I_1}{\partial x}\right|, \left|\frac{\partial I_1}{\partial y}\right|\right]\right\} \quad (5)$$

Furthermore, we acquire self-supervision from cyclic constraints, defined as follows. Given a 3DCT scan pair I_i, I_j and field predictions $\mathbf{F}_{i \to j}$ mapping I_i into I_j, we constrain non-occluded voxels to perform a closed loop:

$$\mathcal{L}^C = \sum_{\mathbf{x} \in \mathcal{N}} \left(\mathbf{F}_{i \to j}\left(\mathbf{x}\right) + \mathbf{F}_{j \to i}\left(\mathbf{x}\right)\right)^2 \quad (6)$$

where $\mathcal{N}$ is the set of non-occluded voxel locations. We define our NCC window size to 7, and our ADMM parameters to $\lambda = \rho = 0.1$, $\alpha = 50$ and $T = 1$, i.e. performing two cost update steps as in [13].

Table 1. L2R results. Our method, ranked 7 on the L2R test data, is compared with several DL based challenge finalists. We report competitive results being fully unsupervised and using no lung segmentation.

Method	L2R Valid.		L2R Test			
	TRE	SDlogJ	TRE	TRE30	SDlogJ	RT
Ours-TV-valid	0.06	3.50	-	-	-	-
Ours-unrolled-valid	0.08	**3.42**	-	-	-	-
LapIRN [16]	-	-	1.98	2.95	0.06	10.3
corrField [5,7]	-	-	**1.75**	**2.48**	0.05	**2.91**
PDD-Net [6]	-	-	2.46	3.81	0.04	4.22
Ours-unrolled-test	-	-	**2.26**	**3.01**	0.07	**2.90**

Results. The results on the Learn2Reg data are given in Table 1, and a qualitative example is given in Fig. 2. Training using our unrolled cost achieved reduced error rates on the Learn2Reg validation set, capturing sharper motion boundaries along the rib cage, compared to standard TV regularization. Furthermore, declared a challenge finalist, our method achieved results on par with the challenge winners on the official test data, simply through expanding a model to the 3D domain and improving its gradients during training.

4   Conclusions

We introduced the concept of Cost Unrolling, shifting algorithm unrolling to the cost function, while preserving model architecture. Our method enables improved training of a TV regularized model as a result of more accurate gradients, thanks to its differentiability around its optimum. We have demonstrated the effectiveness of our method in synthetic problems as well as the real-world unsupervised volumetric displacement fields problem. Our unrolled cost achieved superior results in all tested scenarios. We believe that the proposed framework can be applied on top of other model architectures for boosting their results next to non-differentiable optimum solutions.

Acknowledgements. This work is partially funded by the Zimin Institute for Engineering Solutions Advancing Better Lives, the Israeli consortiums for soft robotics and autonomous driving, the Nicholas and Elizabeth Slezak Super Center for Cardiac Research and Biomedical Engineering at Tel Aviv University and TAU Science Data and AI Center.

References

1. Abadi, M., et al.: TensorFlow: large-scale machine learning on heterogeneous systems (2015). https://www.tensorflow.org/

2. Boyd, S., Parikh, N., Chu, E., Peleato, B., Eckstein, J.: Distributed optimization and statistical learning via the alternating direction method of multipliers. Found. Trends Mach. Learn. **3**(1), 1–122 (2011). https://doi.org/10.1561/2200000016
3. Butler, D.J., Wulff, J., Stanley, G.B., Black, M.J.: A naturalistic open source movie for optical flow evaluation. In: Fitzgibbon, A., Lazebnik, S., Perona, P., Sato, Y., Schmid, C. (eds.) ECCV 2012. LNCS, vol. 7577, pp. 611–625. Springer, Heidelberg (2012). https://doi.org/10.1007/978-3-642-33783-3_44
4. Chambolle, A., Lions, P.: Image recovery via total variation minimization and related problems. Numerische Mathematik **76**, 167–188 (1997). https://doi.org/10.1007/s002110050258
5. Hansen, L., Heinrich, M.P.: GraphRegNet: deep graph regularisation networks on sparse keypoints for dense registration of 3D lung CTs. IEEE Trans. Med. Imaging **40**(9), 2246–2257 (2021)
6. Heinrich, M.P.: Closing the gap between deep and conventional image registration using probabilistic dense displacement networks. In: Shen, D., et al. (eds.) MICCAI 2019. LNCS, vol. 11769, pp. 50–58. Springer, Cham (2019). https://doi.org/10.1007/978-3-030-32226-7_6
7. Heinrich, M.P., Handels, H., Simpson, I.J.A.: Estimating large lung motion in COPD patients by symmetric regularised correspondence fields. In: Navab, N., Hornegger, J., Wells, W.M., Frangi, A.F. (eds.) MICCAI 2015. LNCS, vol. 9350, pp. 338–345. Springer, Cham (2015). https://doi.org/10.1007/978-3-319-24571-3_41
8. Hering, A., et al.: Learn2Reg: comprehensive multi-task medical image registration challenge, dataset and evaluation in the era of deep learning (2021)
9. Hermann, S., Werner, R.: TV-L_1-based 3D medical image registration with the census cost function. In: Klette, R., Rivera, M., Satoh, S. (eds.) PSIVT 2013. LNCS, vol. 8333, pp. 149–161. Springer, Heidelberg (2014). https://doi.org/10.1007/978-3-642-53842-1_13
10. Huang, J., Mumford, D.: Statistics of natural images and models. In: Proceedings. 1999 IEEE Computer Society Conference on Computer Vision and Pattern Recognition (Cat. No PR00149), vol. 1, pp. 541–547 (1999). https://doi.org/10.1109/CVPR.1999.786990
11. Huber, P., Wiley, J., InterScience, W.: Robust Statistics. Wiley, New York (1981)
12. Jonschkowski, R., Stone, A., Barron, J.T., Gordon, A., Konolige, K., Angelova, A.: What matters in unsupervised optical flow. arXiv preprint arXiv:2006.04902 (2020)
13. Lifshitz, G., Raviv, D.: Cost function unrolling in unsupervised optical flow (2021)
14. Liu, L., et al.: Learning by analogy: reliable supervision from transformations for unsupervised optical flow estimation. In: IEEE Conference on Computer Vision and Pattern Recognition (CVPR) (2020)
15. Menze, M., Geiger, A.: Object scene flow for autonomous vehicles. In: Conference on Computer Vision and Pattern Recognition (CVPR) (2015)
16. Mok, T.C.W., Chung, A.C.S.: Large deformation diffeomorphic image registration with Laplacian pyramid networks. In: Martel, A.L., et al. (eds.) MICCAI 2020. LNCS, vol. 12263, pp. 211–221. Springer, Cham (2020). https://doi.org/10.1007/978-3-030-59716-0_21
17. Paszke, A., et al.: PyTorch: an imperative style, high-performance deep learning library. In: Wallach, H., Larochelle, H., Beygelzimer, A., d'Alché-Buc, F., Fox, E., Garnett, R. (eds.) Advances in Neural Information Processing Systems, vol. 32, pp. 8024–8035. Curran Associates, Inc. (2019). http://papers.neurips.cc/paper/9015-pytorch-an-imperative-style-high-performance-deep-learning-library.pdf

18. Rudin, L.I., Osher, S., Fatemi, E.: Nonlinear total variation based noise removal algorithms. Physica D **60**(1), 259–268 (1992). https://doi.org/10.1016/0167-2789(92)90242-F, https://www.sciencedirect.com/science/article/pii/016727899290242F

19. Urschler, M., Werlberger, M., Scheurer, E., Bischof, H.: Robust optical flow based deformable registration of thoracic CT images. In: Medical Image Analysis for the Clinic: A Grand Challenge, pp. 195–204 (2010)

20. Zach, C., Pock, T., Bischof, H.: A duality based approach for realtime TV-L^1 optical flow. In: Hamprecht, F.A., Schnörr, C., Jähne, B. (eds.) DAGM 2007. LNCS, vol. 4713, pp. 214–223. Springer, Heidelberg (2007). https://doi.org/10.1007/978-3-540-74936-3_22

Conditional Deep Laplacian Pyramid Image Registration Network in Learn2Reg Challenge

Tony C. W. Mok[✉] and Albert C. S. Chung

Department of Computer Science and Engineering, The Hong Kong University of Science and Technology, Sai Kung, Hong Kong
{cwmokab,achung}@cse.ust.hk

Abstract. Hyperparameter tuning is extremely tedious and costly in the deep learning-based deformable image registration methods. In this paper, we present our contribution to the Learn2Reg challenge and demonstrate how hyperparameter tuning can be accelerated and simplified with the proposed conditional image registration framework. We exemplify the conditional image registration framework with the deep Laplacian pyramid image registration network (cLapIRN) and apply it comprehensively to all three tasks in the challenge. Our method was ranked the first place in the Learn2Reg 2021 challenge.

Keywords: Image registration · Hyperparameter tuning · Learn2Reg challenge

1 Introduction

Medical image registration is essential in a variety of medical image analysis tasks. Image registration aligns a pair of fixed and moving images with maximizing the spatial similarity between the transformed moving image and the fixed image. Recently, a variety of deep learning-based image registration methods [7,10,16,19] have been developed and proposed to circumvent the tedious iterative optimization process in the conventional image registration methods. While deep learning-based methods are starting to show promising registration performance and speed, hyperparameter tuning in deep learning-based methods remains a challenge. To address this issue, we propose a conditional image registration framework [18] dedicated to enabling rapid hyperparameter tuning in a deep learning-based method. To further ensure comparability and compare to the state-of-the-art image registration algorithms, we have applied our method to the Learn2Reg: 2021 MICCAI Registration Challenge [1,11]. It consists of three different registration tasks: lung CT expiration-inspiration registration [12], thorax-abdomen CT-MR registration [4,14,20], and brain MR inter-patient registration [13,15]. In this paper, we present our solutions to the three tasks in the challenge.

M. Aubreville et al. (Eds.): MIDOG 2021/MOOD 2021/L2R 2021, LNCS 13166, pp. 161–167, 2022.
https://doi.org/10.1007/978-3-030-97281-3_23

2   Methods

2.1   Conditional Deformable Image Registration

Given a fixed image F, a moving 3D image volume M, deep learning-based deformable image registration methods parameterize the image registration problem as a function $f_\theta(F, M, c) = \phi$, where f_θ is the convolutional neural network (CNN). Our conditional deformable image registration framework [18] further extends this formulation to take the hyperparameter of smoothness regularization λ as input, i.e., $f_\theta(F, M, \lambda) = \phi$. The proposed conditional deformable image registration framework aims to learn the effect of the hyperparameter of smoothness regularization on the output deformation field. To condition a CNN model on a conditional variable, we utilize a recently proposed conditional instance normalization (CIN) module [5,18] to replace the high-level layers in the CNN model. Originally, the CIN layer used instance normalization to normalize the data distribution of the feature map before shifting the feature statistics. We found that the instance normalization can be pruned, and the CIN module can still capture the hyperparameter effect. Formally, the simplified CIN operation for each feature map h_i is defined as

$$h_i' = \gamma_{\theta,i}(z)h_i + \beta_{\theta,i}(z), \tag{1}$$

where $\gamma_{\theta,i}, \beta_{\theta,i} \in \mathbb{R}$ denote learning parameters learned from the latent code z.

2.2   Deep Laplacian Pyramid Image Registration Network

All three tasks are comprehensively tackled by a deep learning-based deformable image registration method as the backbone network. Specifically, we modify the deep Laplacian pyramid image registration network (LapIRN) [19] by replacing all the residual blocks with the proposed conditional image registration module (denoted as cLapIRN [18]). LapIRN is particularly well suited for all tasks in the challenge because of its versatile design and promising registration accuracy under large deformation settings. LapIRN utilizes a multi-level, coarse-to-fine CNN architecture to mimic the multi-resolution optimization strategy in the conventional image registration method. Since registration accuracy is more favourable over the diffeomorphic properties, we parameterize the LapIRN model with displacement vector fields. During training, with reference to [13], our method learns to minimize the following objective function:

$$\phi^* = \arg \min_\phi (1 - \lambda_p)\mathcal{L}_{sim}(F, M(\phi)) + \lambda_p \mathcal{L}_{reg}(\phi), \tag{2}$$

where ϕ^* denotes the optimal displacement field ϕ, $\mathcal{L}_{sim}(\cdot, \cdot)$ denotes the dissimilarity function, $\mathcal{L}_{reg}(\cdot)$ represents the smoothness regularization function and λ_p is uniformly sampled over $[0, 1]$. We use a diffusion regularizer on the spatial gradients of displacement fields, i.e., $\mathcal{L}_{reg}(\phi) = ||\nabla\phi||_2^2$. The dissimilarity functions $\mathcal{L}_{sim}$ utilized in this work are different between the three tasks in the challenge,

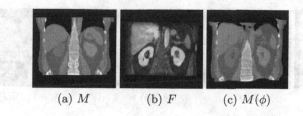

(a) M (b) F (c) $M(\phi)$

Fig. 1. Example coronal slices from a exemplary case in the validation set for task 1: a) the moving CT scan, b) the fixed MR scan and c) the transformed moving scan. Each scan is overlaid with the corresponding anatomical segmentation maps.

which will be discussed in the following sections. The model is trained using a progressive training scheme in a coarse-to-fine manner. The pyramid level and the number of conditional image registration modules for cLapIRN are set to 3 and 5 for all tasks, respectively. For all tasks, we manually balance the trade-off between registration accuracy and the plausibility of the deformation by varying the lambda and grid search with step size 0.01 on the validation set provided by the challenge. For more details of the conditional image registration framework, we recommend the readers refer to our previous works [18,19].

2.3   Task 1 - Thorax-Abdomen CT-MR Registration

The goal of the first task is related to the thorax-abdomen CT-MR intra-patient registration. Our submitted method combines a modality-independent neighborhood descriptor (MIND-SSC) [8,9] with cLapIRN. First, we extract the MIND-SSC features of the fixed image and moving image. Then, cLapIRN take as input the MIND-SSC features of the fixed and moving images. We use the mean square error (MSE) and the Dice coefficient of the provided anatomical segmentation map as dissimilarity functions. Due to the limited GPU memory, we further downsample the input data to half-resolution. Yet, the evaluation and similarity measures are computed at full resolution by upsampling the output deformation field with bilinear interpolation. We mask the similarity measure with the region of interest (ROI) mask of the fixed image. To alleviate the overfitting issue as well as fully utilize the unpaired data, we first train our model with all training data and treat it as an inter-patient registration problem. Then, we finetune our model on the 8 paired training data (5 and 3 pairs for training and validation, respectively). Throughout the training, a fast affine augmentation is adopted to diversify the training data further.

2.4   Task 2 - Lung CT Expiration-Inspiration Registration

The aim of the second task is related to lung CT expiration-inspiration registration. We utilize the vanilla cLapIRN with instance optimization as postprocessing to address the challenges in this task. We use the local normalized cross-correlation (NCC) similarity measure with the similarity pyramid for training

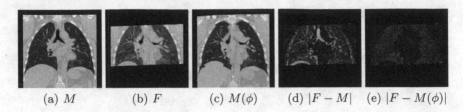

(a) M (b) F (c) $M(\phi)$ (d) $|F - M|$ (e) $|F - M(\phi)|$

Fig. 2. Example coronal slices from a exemplary case in the validation set for task 2: a) the expiration CT scan, b) the inspiration CT scan, c) the registered inspiration CT scan, d) the difference image before the registration and e) the difference image after registration. For better visualization, we masked the difference images with the ROI of the expiration scan.

and the normalized gradient fields (NGF) [6] with $\epsilon = 0.1$ as distance measure for the instance optimization. Although massive pseudo landmarks are provided, we use the pseudo landmarks for validation only. Similar to task 1, we use affine augmentation to alleviate the overfitting issue during training. We also mask the NCC and NGF similarity measure with the expiration lung mask. For the instance optimization, we use the output deformation field from cLapIRN as initialization and utilize Adam optimizer with a learning rate of 0.005 to minimize the similarity function and the weighted smoothness regularization for 60 iterations.

2.5   Task 3 - Brain MR Inter-patient Registration

The objective of the third task is brain MR inter-patient registration. Since the main source of misalignment in the provided data is non-linear, our submitted method uses the vanilla cLapIRN to address the challenge in this task. We use the NCC similarity measure with the similarity pyramid and the Dice score of the subcortical structures segmentation map as similarity measures, as shown in our previous work [17]. Unlike previous tasks, our submitted method does not use affine augmentation during training, as we observe that affine augmentation does not improve the registration performance in the validation set.

3   Results

The Learn2Reg challenge computes a sequence of metrics, including TRE, TRE30, DSC, DSC30, HD95 and $std(|J_\phi|)$, in order to provide a comprehensive evaluation of the registration performance. Specifically, DSC and HD95 are the dice similarity coefficient of segmentations and 95% percentile of Hausdorff distance of segmentations, which quantify the registration accuracy for tasks 1 and 3. TRE denotes the target registration error of manual landmarks, which is used to measure the registration accuracy for task 2. The plausibility of the solutions is measured using the standard deviation of log Jacobian determinant of

Table 1. Results of our method on the validation set in the Learn2Reg challenge. More results can be found at the validation leaderboard of the challenge [2].

Method	Task 1					Task 2			Task 3				
	DSC	DSC30	HD95	$std(|J_\phi|)$	T_{test}	TRE	$std(|J_\phi|)$	T_{test}	DSC	DSC30	HD95	$std(|J_\phi|)$	T_{test}
Initial	0.30	0.23	22.89	-	-	14.64	-	-	0.57	0.55	3.83	-	-
Ours	0.90	0.90	2.75	0.08	0.35	2.13	0.07	8.21	0.86	0.86	1.51	0.07	0.24

Table 2. Results of our method on the test set in the Learn2Reg challenge. The comprehensive results of the other participants are shown on the official website of the Learn2Reg Challenge [3].

Method	Task 1					Task 2				Task 3				
	DSC	DSC30	HD95	$std(|J_\phi|)$	T_{test}	TRE	TRE30	$std(|J_\phi|)$	T_{test}	DSC	DSC30	HD95	$std(|J_\phi|)$	T_{test}
Ours	0.76	0.69	18.6	0.12	1.5	2.0	3.0	0.06	10.3	0.82	0.68	1.98	0.07	1.2

the deformation field (denoted as $std(|J_\phi|)$ for all tasks. DSC30 and TRE30 are computed by averaging the lowest 30% of the lowest DSC and TRE, respectively. Furthermore, the registration time per case, excluding the GPU initialization, is measured in the unit of seconds and denoted as T_{test} in Tables 1 and 2.

Tables 1 and 2 summarize our results in the validation set and test set of the challenge. Figures 1 and 2 show the qualitative results of tasks 1 and 2, respectively. While our method is able to achieve remarkable results in the validation set of tasks 1 and 3 (improving the DSC from 0.30 to 0.90 and from 0.57 to 0.86, respectively), we observe a significant drop in registration performance in tasks 1 and 3. The degraded registrations could be the result of overfitting to the particular imaging protocol, and the generalization ability of our model needs further investigation. Yet, with a mean DSC of 0.76 and 0.82 along with the 1.5 and 1.2 average inference time in tasks 1 and 3, respectively, our method can achieve the best overall result in task 3 and the second place in task 1. Moreover, our method achieves an average TRE of 2.13 and 2.0 in the validation and test set of task 2, suggesting that our method is capable of generating comparable results with the conventional image registration approaches with a small amount of training data.

4   Conclusion

We have successfully applied the cLapIRN to all three tasks in the Learn2Reg challenge. Our method can achieve promising results among all tasks, including intra- and inter-patient registration tasks, cross-modalities and anatomies registration tasks, implying that cLapIRN is a generic and versatile deep learning-based method. Our method has achieved the first place in the challenge. Furthermore, not only did we obtain the best overall result in task 3, we have demonstrated that our proposed method is capable of generating comparable solutions with the conventional image registration methods even in tasks with a small training dataset, i.e., 20 paired CT scans in task 2.

References

1. Learn2Reg: 2021 MICCAI registration challenge. https://learn2reg.grand-challenge.org/. Accessed 8 Oct 2021
2. Validation leaderboard of Learn2Reg 2021 MICCAI registration challenge. https://learn2reg.grand-challenge.org/evaluation/test/leaderboard/. Accessed 8 Oct 2021
3. Workshop schedule of Learn2Reg 2021 MICCAI registration challenge. https://learn2reg.grand-challenge.org/Workshop/. Accessed 8 Oct 2021
4. Clark, K., et al.: The Cancer Imaging Archive (TCIA): maintaining and operating a public information repository. J. Digit. Imaging 26(6), 1045–1057 (2013). https://doi.org/10.1007/s10278-013-9622-7
5. Dumoulin, V., Shlens, J., Kudlur, M.: A learned representation for artistic style. International Conference on Learning Representations (2017)
6. Haber, E., Modersitzki, J.: Intensity gradient based registration and fusion of multi-modal images. In: Larsen, R., Nielsen, M., Sporring, J. (eds.) MICCAI 2006. LNCS, vol. 4191, pp. 726–733. Springer, Heidelberg (2006). https://doi.org/10.1007/11866763_89
7. Heinrich, M.P.: Closing the gap between deep and conventional image registration using probabilistic dense displacement networks. In: Shen, D., et al. (eds.) MICCAI 2019. LNCS, vol. 11769, pp. 50–58. Springer, Cham (2019). https://doi.org/10.1007/978-3-030-32226-7_6
8. Heinrich, M.P., et al.: MIND: modality independent neighbourhood descriptor for multi-modal deformable registration. Med. Image Anal. 16(7), 1423–1435 (2012)
9. Heinrich, M.P., Jenkinson, M., Papież, B.W., Brady, S.M., Schnabel, J.A.: Towards realtime multimodal fusion for image-guided interventions using self-similarities. In: Mori, K., Sakuma, I., Sato, Y., Barillot, C., Navab, N. (eds.) MICCAI 2013. LNCS, vol. 8149, pp. 187–194. Springer, Heidelberg (2013). https://doi.org/10.1007/978-3-642-40811-3_24
10. Hering, A., Häger, S., Moltz, J., Lessmann, N., Heldmann, S., van Ginneken, B.: CNN-based lung CT registration with multiple anatomical constraints. Med. Image Anal. 72, 102139 (2021)
11. Hering, A., et al.: Learn2Reg: comprehensive multi-task medical image registration challenge, dataset and evaluation in the era of deep learning. arXiv preprint arXiv:2112.04489 (2021)
12. Hering, A., Murphy, K., van Ginneken, B.: Lean2Reg challenge: CT lung registration-training data (2020)
13. Hoopes, A., Hoffmann, M., Fischl, B., Guttag, J., Dalca, A.V.: HyperMorph: amortized hyperparameter learning for image registration. In: Feragen, A., Sommer, S., Schnabel, J., Nielsen, M. (eds.) IPMI 2021. LNCS, vol. 12729, pp. 3–17. Springer, Cham (2021). https://doi.org/10.1007/978-3-030-78191-0_1
14. Kavur, A.E., et al.: CHAOS challenge-combined (CT-MR) healthy abdominal organ segmentation. Med. Image Anal. 69, 101950 (2021)
15. Marcus, D.S., Wang, T.H., Parker, J., et al.: Open Access Series of Imaging Studies (OASIS): cross-sectional MRI data in young, middle aged, nondemented, and demented older adults. J. Cogn. Neurosci. 19(9), 1498–1507 (2007)

16. Mok, T.C., Chung, A.: Fast symmetric diffeomorphic image registration with convolutional neural networks. In: CVPR, pp. 4644–4653 (2020)
17. Mok, T.C.W., Chung, A.C.S.: Large deformation image registration with anatomy-aware Laplacian pyramid networks. In: Shusharina, N., Heinrich, M.P., Huang, R. (eds.) MICCAI 2020. LNCS, vol. 12587, pp. 61–67. Springer, Cham (2021). https://doi.org/10.1007/978-3-030-71827-5_7
18. Mok, T.C.W., Chung, A.C.S.: Conditional deformable image registration with convolutional neural network. In: de Bruijne, M., et al. (eds.) MICCAI 2021. LNCS, vol. 12904, pp. 35–45. Springer, Cham (2021). https://doi.org/10.1007/978-3-030-87202-1_4
19. Mok, T.C.W., Chung, A.C.S.: Large deformation diffeomorphic image registration with Laplacian pyramid networks. In: Martel, A.L., et al. (eds.) MICCAI 2020. LNCS, vol. 12263, pp. 211–221. Springer, Cham (2020). https://doi.org/10.1007/978-3-030-59716-0_21
20. Xu, Z., Lee, C.P., Heinrich, M.P., et al.: Evaluation of six registration methods for the human abdomen on clinically acquired CT. IEEE Trans. Biomed. Eng. 63(8), 1563–1572 (2016)

The Learn2Reg 2021 MICCAI Grand Challenge (PIMed Team)

Wei Shao[1]([✉]), Sulaiman Vesal[2]([✉]), David Lim[3], Cynthia Li[4], Negar Golestani[1], Ahmed Alsinan[1], Richard Fan[2], Geoffrey Sonn[2], and Mirabela Rusu[1]([✉])

[1] Department of Radiology, Stanford University, Stanford, CA 94305, USA
{weishao,mirabela.rusu}@stanford.edu
[2] Department of Urology, Stanford University, Stanford, CA 94305, USA
svesal@stanford.edu
[3] Department of Computer Science, Stanford University, Stanford, CA 94305, USA
[4] Institute for Computational and Mathematical Engineering, Stanford University, Stanford, CA 94305, USA

Abstract. This paper summarizes our approaches and results for the three tasks of the Learn2Reg 2021 MICCAI Grand Challenge focused on the registration of: (1) intra-patient abdominal CT and MR images, (2) intra-patient expiration and inspiration lung CT scans, and (3) inter-patient brain MR images. These registration tasks have multiple challenges including dealing with multi-modal scans, estimating large deformations, lack of training data, and missing data. For Task 1, we first segmented four organs in both CT and MRI and, second, used them in a two-stage deformable registration pipeline. Our approach has achieved a Dice coefficient of 0.71. For Task 2, we handled missing data in the expiration CT by using a pairwise geodesic density registration algorithm that excludes data outside the lungs. Our approach has achieved a target registration error of 2.3 mm. For Task 3, we modified the Voxel-Morph architecture to give more degrees of freedom to the registration model and used it to register brain MRI across patients. Our approach has achieved a Dice coefficient of 0.78. Overall, our team has won second place out of 35 submissions from 15 teams.

Keywords: Image registration · Diffeomorphic registration · Deep learning image registration · CT · MRI

1 Introduction

The goal of image registration is to find a geometric transformation that defines the point-to-point correspondences between two images. Image registration is a challenging task due to multiple modalities, lack of training dataset, large displacement deformation, missing information, alignment of small structures, and inter-patient alignment. The Learn2Reg 2021 MICCAI Grand Challenge [4]

W. Shao and S. Vesal—Equal contribution as first authors.
G. Sonn and M. Rusu—Equal contribution as senior authors.

M. Aubreville et al. (Eds.): MIDOG 2021/MOOD 2021/L2R 2021, LNCS 13166, pp. 168–173, 2022.
https://doi.org/10.1007/978-3-030-97281-3_24

provides an accessible medical image registration benchmark for comprehensive evaluation of traditional and deep learning registration methods on three tasks that cover the above mentioned challenges. The first task is the multi-modal registration of CT and MR images of the abdomen. The second task is to estimate the large displacement deformation between the expiration and inspiration CT scans. The third task is the inter-patient registration of brain MR images. For each task, we have developed a customized approach using either traditional and/or deep learning registration. Our approaches are summarized below.

2   Methods

2.1   Dataset

For Task 1, paired abdomen CT and MR images of five, three and eight patients from The Cancer Imaging Archive (TCIA) [2] were provided for the training, validation, and testing, respectively. Additional unpaired 50 CT and 40 MR images were also provided for the training. Manual segmentation of the liver, spleen, left kidney, and right kidney on the CT and MR images were also provided for the training. For Task 2, 20, three, and 10 paired expiration-inspiration CT scans were provided for the training, validation, and testing, respectively. Binary lung segmentations were also provided for both training and testing. For Task 3, skull-stripped 3D MR scans of 400, 19, and 38 patients were provided for the training, validation, and testing, respectively. The images were normalized and cropped to $160 \times 192 \times 224$ volumes. The segmentation masks for four brain anatomical regions (white matter, gray matter, hippocampi and ventricles) were also provided for training and validation set only.

2.2   Task 1: Intra-patient CT-MR Registration

We tackled the multi-modal CT-MR image registration problem in two steps (Fig. 1). First, we segmented the liver, spleen, left kidney, and right kidney on both CT and MR images by training three 2D U-Net segmentation networks [6] with 2D slices from the axial, coronal, and sagittal views. We augmented the data by applying random affine and elastic deformations, cropping, flipping, blurring, and Gaussian-noise to the training images, as well as varying the brightness and contrast of the training images. The average of the three views produced the final 3D segmentations on CT and MR images. We trained using the Adam optimizer, a learning rate of 10^{-4} for the first 200 epochs and a learning rate of 10^{-5} for the remaining 100 epochs. Second, we registered the MRI and CT in two steps. For the first step, we estimate a rigid transformation between the two images by using CT and MR segmentations of the liver, spleen, left kidney, and right kidney. For second step, we estimated a diffeomorphic transformation using a segmentation loss (sum of squared differences) and an intensity loss (normalized cross correlation). If all four organs were segmented, we gave a weight of 0.75 to the segmentation loss and a weight of 0.25 to the intensity loss . If not all the organs were segmented due to pathologies or poor data quality, we gave equal weight of 0.5 to the segmentation and the intensity losses.

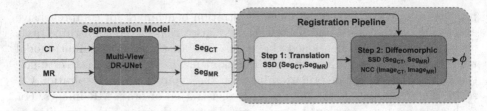

Fig. 1. The framework for multi-modal CT-MR registration, which includes a segmentation network and a subsequent deformable registration model.

2.3 Task 2: Intra-patient Expiration-Inspiration CT Registration

We registered the expiration-inspiration CT scans with a pairwise geodesic density registration algorithm [5] that is robust to missing information in the expiration CT by using a binary mask to exclude data outside the lungs from the registration. The binary mask indicated the locations that do not belong to the lungs. Our geodesic registration algorithm also considers changes in CT intensity that are associated with local lung volume change. We first convert each CT image in the Hounsfield unit into a tissue density image via

$$I = \frac{CT - HU_{air}}{HU_{tissue} - HU_{air}} = \frac{CT + 1000}{1055} \tag{1}$$

The intensity of a tissue density image ranges from 0 to 1, where 0 corresponds to 0% lung tissue at a voxel and 1 corresponds to 100% lung tissue at a voxel. Our geodesic density registration algorithm uses tissue density images instead of CT intensity images for the registration. We used the following tissue density deformation action to deform a tissue density image by a transformation.

$$\phi \cdot I := |D\phi^{-1}| I \circ \phi^{-1} \tag{2}$$

Here, the Jacobian determinant of the transformation accounts for tissue density change associated with local lung volume change. The inputs to our geodesic density registration algorithm are two tissue density images and the corresponding binary artifact masks. The output of the algorithm is a time-varying velocity field v_t that parameterizes a flow of diffeomorphisms ϕ_t to deform the image I_0 into a geodesic image flow. The cost function consists of two terms.

$$E(v_t) = \underbrace{\frac{1}{2} \int_0^1 <v_t, v_t>_{\mathfrak{g}} dt}_{\text{smoothness}} + \underbrace{\frac{1}{\sigma^2}||(\phi_1 \cdot I_0 - I_1)M_1||_{L_2}^2}_{\text{image difference}} \tag{3}$$

The first term measures the smoothness of the transformation, and the second term measures the difference between the predicted image flow and the input tissue density image at time 1. Notice that the difference image in the second term is multiplied by a binary mask M_1 that indicates the location of the lungs, so that image differences were only computed for voxels inside the lungs.

Table 1. Summary of registration results on testing datasets for all three tasks.

Task	Dice	TRE (mm)	HD95 (mm)	SDLogJ	Run time (s)
Task 1	0.71	N/A	14.2	0.07	59
Task 2	N/A	2.34	NA	0.04	623
Task 3	0.78	N/A	1.84	0.06	5.9

2.4 Task 3: Inter-patient Brain MRI Registration

Prior VoxelMorph [1] network is a state-of-the-art deep learning approach for brain MR image registration. Inspired by this approach, we replaced the regular U-Net with the residual U-Net to give more complexity to the network. We trained our network by using the following cost function, which is a weighted Sum of the Sum of Squared Differences loss $\mathcal{L}_{SSD}$, a regularization loss $\mathcal{L}_{smooth}$ and a multi-class weighted dice loss $\mathcal{L}_{Dice}$.

$$\mathcal{L} = \lambda_1 \mathcal{L}_{SSD} + \lambda_2 \mathcal{L}_{smooth} + \lambda_3 \mathcal{L}_{Dice} \tag{4}$$

where $\lambda_1 = 1.0$, $\lambda_1 = 0.01$ and $\lambda_3 = 0.05$ are the weights controlling the effect of each term. We augmented the data by applying horizontal flipping and contrast enhancement. We trained our network by using the Adam optimizer, a learning rate of 10^{-4} for the first 1500 epochs and 10^{-5} for the rest 500 epochs.

2.5 Evaluation Metrics

We utilized several metrics provided by the challenge organizer for each task to evaluate the performance of our registration methods. These metrics include the Target Registration Error (TRE) of landmarks, Dice Similarity Coefficient (DSC), Hausdorff Distance (HD95) of segmentations, and standard deviation of log Jacobian determinant (SDlogJ) of the deformation field [3]. Moreover, we reported the run time of each registration method on test patients.

3 Results and Discussion

Table 1 summarizes our results on the test dataset. For Task 1, the two-step registration method achieved a DSC score of 0.71, HD95 of 14.2 mm, and SDLogJ of 0.07 with a run time of 59 s. For Task 2, our pairwise geodesic registration method achieved a TRE of 2.34 mm and SDLogJ of 0.04. The run time for this task was 623 s, due to using a traditional registration approach. For Task 3, our residual registration model obtained a Dice score of 0.78, HD95 of 1.84 mm, and SDLogJ of 0.06, with a run-time of 5.9 s.

Figure 2, 3 and 4 show the registration results of one representative subject for each task. As shown in Fig. 2, despite the fact that the CT and MR images are significantly different in terms of contrast and anatomical organ appearance,

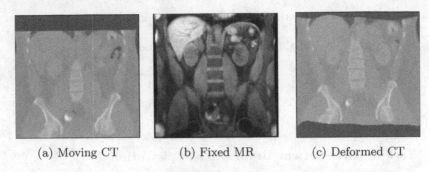

(a) Moving CT   (b) Fixed MR   (c) Deformed CT

Fig. 2. Example MR (fixed image), CT (moving image), and deformed CT slices for multi-modal CT-MR registration task.

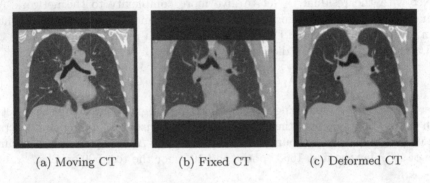

(a) Moving CT   (b) Fixed CT   (c) Deformed CT

Fig. 3. Example lung CT slices of expiration (fixed image), inspiration (moving image), and deformed image.

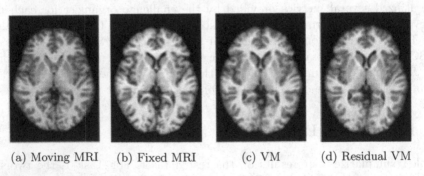

(a) Moving MRI   (b) Fixed MRI   (c) VM   (d) Residual VM

Fig. 4. Example MR slices of input moving image, fixed image, VoxelMorph and our residual VoxelMorph deformed image.

the deformed CT aligned quite well with the MR scan. Figure 3 shows that our PGDR algorithm has accurately estimated the large displacement deformation between the expiration and inspiration CT scans despite of the missing data in the expiration CT scan. For Task 3, the residual VoxelMorph implementation produced smoother transformations for brain MRI images than the conventional VoxelMorph implementation, resulting in a lower SDLogJ score (Fig. 4).

4   Conclusion

In this paper, we presented the results and methods for three different registration tasks in the Learn2Reg 2021 MICCAI Grand Challenge. To address the challenges including missing data, multiple modalities and large deformation in the tasks, we have developed registration methods with a combination of traditional and deep learning approaches. The proposed methods achieved promising registration results evaluated based on various metrics. In the future, we aim to replace the traditional registration components in our approaches with learning-based methods to improve the run-time and model generalization across the different datasets.

Acknowledgments. This work was supported by the Department of Radiology and the Department of Urology at Stanford University.

References

1. Balakrishnan, G., Zhao, A., Sabuncu, M.R., Guttag, J., Dalca, A.V.: VoxelMorph: a learning framework for deformable medical image registration. IEEE Trans. Med. Imaging **38**(8), 1788–1800 (2019)
2. Clark, K., et al.: The Cancer Imaging Archive (TCIA): maintaining and operating a public information repository. J. Digit. Imaging **26**(6), 1045–1057 (2013)
3. Dalca, A.V., Balakrishnan, G., Guttag, J., Sabuncu, M.R.: Unsupervised learning for fast probabilistic diffeomorphic registration. In: Frangi, A.F., Schnabel, J.A., Davatzikos, C., Alberola-López, C., Fichtinger, G. (eds.) MICCAI 2018. LNCS, vol. 11070, pp. 729–738. Springer, Cham (2018). https://doi.org/10.1007/978-3-030-00928-1_82
4. Hering, A., et al.: Learn2Reg: comprehensive multi-task medical image registration challenge, dataset and evaluation in the era of deep learning. arXiv preprint arXiv:2112.04489 (2021)
5. Shao, W., et al.: Geodesic density regression for correcting 4DCT pulmonary respiratory motion artifacts. Med. Image Anal. **72**, 102140 (2021)
6. Vesal, S., Gu, M., Kosti, R., Maier, A., Ravikumar, N.: Adapt everywhere: unsupervised adaptation of point-clouds and entropy minimization for multi-modal cardiac image segmentation. IEEE Trans. Med. Imaging **40**(7), 1838–1851 (2021)

Fast 3D Registration with Accurate Optimisation and Little Learning for Learn2Reg 2021

Hanna Siebert(✉)🆔, Lasse Hansen🆔, and Mattias P. Heinrich🆔

Institute of Medical Informatics, Universität zu Lübeck, Lübeck, Germany
{siebert,hansen,heinrich}@imi.uni-luebeck.de

Abstract. Current approaches for deformable medical image registration often struggle to fulfill all of the following criteria: versatile applicability, small computation or training times, and the being able to estimate large deformations. Furthermore, end-to-end networks for supervised training of registration often become overly complex and difficult to train. For the Learn2Reg2021 challenge, we aim to address these issues by decoupling feature learning and geometric alignment. First, we introduce a new very fast and accurate optimisation method. By using discretised displacements and a coupled convex optimisation procedure, we are able to robustly cope with large deformations. With the help of an Adam-based instance optimisation, we achieve very accurate registration performances and by using regularisation, we obtain smooth and plausible deformation fields. Second, to be versatile for different registration tasks, we extract hand-crafted features that are modality and contrast invariant and complement them with semantic features from a task-specific segmentation U-Net. With our results we were able to achieve the overall Learn2Reg2021 challenge's second place, winning Task 1 and being second and third in the other two tasks.

Keywords: image registration · convex optimisation · instance optimisation

1 Motivation

Deep-learning-based approaches for medical image registration usually involve an elaborate learning procedure and yet they often struggle with the estimation of large deformations and the versatile usability for a wide range of tasks. To address the different registration tasks of the Learn2Reg2021 challenge[1] [8], we present a fast and accurate optimisation method for image registration that requires little learning. Our method robustly captures large deformations by using discretised displacements and a coupled convex optimisation. In order to be versatile for various tasks, we include a hand-crafted feature extractor in our method that is contrast and modality invariant and still highly discriminative for local geometry.

---

[1] https://learn2reg.grand-challenge.org.

© Springer Nature Switzerland AG 2022
M. Aubreville et al. (Eds.): MIDOG 2021/MOOD 2021/L2R 2021, LNCS 13166, pp. 174–179, 2022.
https://doi.org/10.1007/978-3-030-97281-3_25

2   Methods

The main idea of our method is to perform large-deformation image registration by using a coupled convex optimisation [6] that approximates a globally optimal solution of a discretised cost function followed by an Adam-based instance optimisation to further improve the local registration accuracy. Dense correlation has already been used extensively in learning based optical flow estimation (cf. PWC-Net [16]) and end-to-end trainable 3D registration networks (cf. PDD-Net [5]), however both approaches have limitations. PWC-Net requires multiple warping steps and is difficult to extend from 2D to 3D (see [4]). PDD-Net employs a dense 3D displacements, but substantially simplifies the optimisation strategy, which may lead to some inaccuracies. ConvexAdam aims to combine the best of both worlds (learning and optimisation-based) by leveraging segmentation priors where available and relying on robust hand-crafted features and fast discrete optimisation.

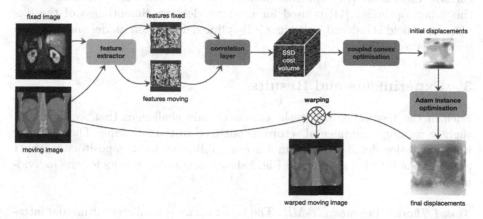

Fig. 1. The structure of our registration method. It consists of a feature extractor (MIND and/or nnUNet) and a dense correlation layer followed by a coupled convex optimisation and an Adam-based instance optimisation.

As visualised in Fig. 1, the basic structure of our registration method consists of a feature extractor, a correlation layer, a coupled convex optimisation, and an instance optimisation.

The feature extractor outputs contrast and modality invariant features from the fixed and moving input images. For this, hand-crafted MIND features [7] ensuring versatility regarding different types of registration tasks can be employed. Depending on the availability of labelled image data, automatic segmentations as provided by the nnU-Net [10] can be used instead. Different to other state-of-the-art supervised deep learning registration methods [14] we avoid using the expert labels only at the end for the warping loss, which may lead to sub-optimal results due to limited gradient backflow. We instead found that using off-the-shelf segmentation networks produce best results.

The obtained features are fed into a correlation layer, which computes a sum-of-squared-differences (SSD) cost volume with a box filter and gives an initial best displacement for each voxel (simply taking the argmin). Therefore, we employ a search space with up to 5000 discretised displacements per voxel. The capture range can be up to at least 48 voxels in each dimension (setting for Task 2) and therefore estimate large motion accurately.

The correlation layer's output is used to solve two coupled convex optimisation problems for efficient global regularisation: In several iterations, alternating steps are performed for similarity and smoothness optimisation, i.e. a spatially smoothed field based on the current argmin (minimal SSD costs) displacements followed the by adding a penalty to the discreted SSD costs based on the discrepancy of this current globally smooth optimum.

The resulting displacements in turn are used as a starting point for an Adam-based instance optimisation in order to provide the final deformation grid used for warping of the moving input image. This step is very similar to classic optical flow estimation [15]. For this purpose, the cost function is linearised and the Adam optimiser [11] is used for gradient descent. Smoothness of the displacement field is induced by adding a B-spline deformation model and diffusion regularisation.

3   Experiments and Results

Each of the Learn2Reg2021 tasks entails certain challenges that we face with slightly varying experimental setups as outlined in the following. The complete implementation details can be found in our publicly available repository.[2] Table 1 presents quantitative results and Fig. 2 shows qualitative results for the individual tasks.

Task 1 Thorax-Abdomen CT-MR. The first task aims to align multimodal intra-patient data [1–3,12]. Besides of multimodal image registration, the objectives of learning from few and noisy labels, as well as dealing with large deformations and missing correspondences are challenging. For this task, we extract hand-crafted MIND features and include an inverse-consistency constraint as introduced in [6] to enforce a minimised discrepancy between the forward and backward transformations in order to avoid implausible deformations. To further regularise the displacement field during Adam instance optimisation, we add thin plate splines yielding smooth deformation fields. As large deformations are to be expected, we chose a search space that includes discretised displacements with a capture range of 64 mm for each dimension within the scanned anatomy.

Task 2 Lung CT. The second task is to perform inspiration-expiration registration on intra-patient lung CT data [9]. In this task, there is the challenge of estimating large breathing motion for scans with only partial visibility of the lungs in the expiration scans. The displacement search range is selected in order

---

[2] https://github.com/multimodallearning/convexAdam.

Table 1. Results for the different Learn2Reg2021 tasks. Accuracy is measured by the Dice similarity of organ segmentations (Dice), the target registration error for anatomical landmarks (TRE), and the 95% Hausdorff distance for segmentations (HD). Robustness is measured by the 30% lowest Dice scores ($Dice_{30}$), Dice scores for additional segmentations ($Dice_{+add}$) and the 30% highest TRE values (TRE_{30}). Plausibility of the deformations is measured by the standard deviation of the logarithmic Jacobian determinant (SDlogJ). Dice similarities are reported in %, TRE and HD values are given in millimetres and inference time is given in seconds. The last table displays the challenge scores and ranks for the overall 1st, 2nd, and 3rd place.

Task 1	Dice	$Dice_{+9}$	HD	SDlogJ	time
initial	33.1	22.3	44.48	–	–
ours	75.4	73.1	20.75	0.09	1.30

Task 2	TRE	TRE_{30}	SDlogJ	time
initial	10.24	16.80	-	-
ours	1.85	2.89	0.06	1.82

Task 3	Dice	$Dice_{30}$	HD	SDlogJ	time
initial	55.9	29.7	4.07	-	-
ours	79.9	64.5	2.00	0.05	12.62

Scores and ranks	Task1	Task2	Task3
	score (rank)	score (rank)	score (rank)
LapIRN	0.86 (2)	0.79 (4)	0.94 (1)
convexAdam	0.88 (1)	0.83 (3)	0.82 (2)
PIMed	0.85 (4)	0.68 (6)	0.70 (5)

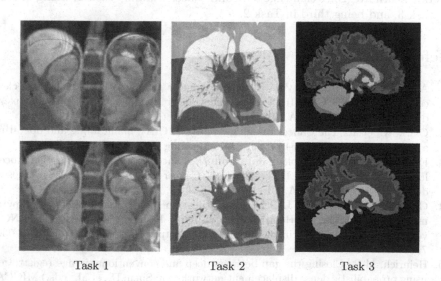

Task 1   Task 2   Task 3

Fig. 2. Qualitative results of our proposed method (top row: colourmap overlay of fixed and moving image (Task 1 and 2) or segmentation (Task 3); bottom row: overlay of fixed and warped moving image or segmentation).

to capture motion with up to $42 \times 30 \times 42$ mm for the x-, y-, and z-dimension respectively. Like in the first task, MIND features of both input images are used to compute the SSD cost volume.

Task 3 Whole Brain MR. The third task deals with the registration of inter-patient T1-weighted brain MRI [13]. Here, the main challenge is to precisely align small structures of variable shape. For this reason, we chose a displacement capture range of 16 mm for each dimension within the scanned brain structures. As this task comprises a large amount of labelled image data, nnU-Net predictions for segmentation guidance are employed. We use the nnU-Net predictions in the form of inverse class-weighted one-hot encodings as features for our method's optimisation steps.

4   Conclusion

Our contribution to the Learn2Reg2021 challenge showed that image registration can be performed fast and accurately using an optimisation strategy with little learning. It is highly parallelisable on a GPU and robust by using a large search space of discretised displacements. Smoothness of the deformation fields could be induced by a global convex regularisation, diffusion regularisation, and B-spline interpolation. By using an efficient Adam-based instance optimisation, our method yields very precise results and by integrating a modality-invariant feature extractor, we achieve a wide versatility. We were able to achieve the overall Learn2Reg2021 challenge's second place, winning Task 1, being second in Task 3, and being third in Task 2.

References

1. Akin, O., et al.: Radiology data from the cancer genome atlas kidney renal clear cell carcinoma [TCGA-KIRC] collection. Cancer Imaging Arch. (2016). https://doi.org/10.7937/K9/TCIA.2016.V6PBVTDR
2. Clark, K., et al.: The Cancer Imaging Archive (TCIA): maintaining and operating a public information repository. J. Digit. Imaging **26**(6), 1045–1057 (2013)
3. Erickson, B., et al.: Radiology data from the cancer genome atlas liver hepatocel-lular carcinoma [TCGA-LIHC] collection. Cancer Imaging Arch. (2016). https://doi.org/10.7937/K9/TCIA.2016.IMMQW8UQ
4. Gunnarsson, N., Sjölund, J., Schön, T.B.: Learning a deformable registration pyra-mid. In: Shusharina, N., Heinrich, M.P., Huang, R. (eds.) MICCAI 2020. LNCS, vol. 12587, pp. 80–86. Springer, Cham (2021). https://doi.org/10.1007/978-3-030-71827-5_10
5. Heinrich, M.P.: Closing the gap between deep and conventional image registration using probabilistic dense displacement networks. In: Shen, D., et al. (eds.) MICCAI 2019. LNCS, vol. 11769, pp. 50–58. Springer, Cham (2019). https://doi.org/10.1007/978-3-030-32226-7_6

6. Heinrich, M.P., Papież, B.W., Schnabel, J.A., Handels, H.: Non-parametric discrete registration with convex optimisation. In: Ourselin, S., Modat, M. (eds.) WBIR 2014. LNCS, vol. 8545, pp. 51–61. Springer, Cham (2014). https://doi.org/10.1007/978-3-319-08554-8_6

7. Heinrich, M.P., Jenkinson, M., Papież, B.W., Brady, S.M., Schnabel, J.A.: Towards realtime multimodal fusion for image-guided interventions using self-similarities. In: Mori, K., Sakuma, I., Sato, Y., Barillot, C., Navab, N. (eds.) MICCAI 2013. LNCS, vol. 8149, pp. 187–194. Springer, Heidelberg (2013). https://doi.org/10.1007/978-3-642-40811-3_24

8. Hering, A., et al.: Learn2Reg: comprehensive multi-task medical image registration challenge, dataset and evaluation in the era of deep learning (2021)

9. Hering, A., Murphy, K., van Ginneken, B.: Learn2Reg challenge: CT lung registration - training data, May 2020. https://doi.org/10.5281/zenodo.3835682

10. Isensee, F., Jaeger, P.F., Kohl, S.A., Petersen, J., Maier-Hein, K.H.: nnU-Net: a self-configuring method for deep learning-based biomedical image segmentation. Nat. Meth. 18(2), 203–211 (2021)

11. Kingma, D.P., Ba, J.: Adam: a method for stochastic optimization. arXiv preprint arXiv:1412.6980 (2014)

12. Linehan, M., et al.: Radiology data from the cancer genome atlas cervical kidney renal papillary cell carcinoma [KIRP] collection. Cancer Imaging Arch. (2016). https://doi.org/10.7937/K9/TCIA.2016.ACWOGBEF

13. Marcus, D.S., Wang, T.H., Parker, J., Csernansky, J.G., Morris, J.C., Buckner, R.L.: Open Access Series of Imaging Studies (OASIS): cross-sectional MRI data in young, middle aged, nondemented, and demented older adults. J. Cogn. Neurosci. 19(9), 1498–1507 (2007)

14. Mok, T.C.W., Chung, A.C.S.: Large deformation image registration with anatomy-aware Laplacian pyramid networks. In: Shusharina, N., Heinrich, M.P., Huang, R. (eds.) MICCAI 2020. LNCS, vol. 12587, pp. 61–67. Springer, Cham (2021). https://doi.org/10.1007/978-3-030-71827-5_7

15. Papenberg, N., Bruhn, A., Brox, T., Didas, S., Weickert, J.: Highly accurate optic flow computation with theoretically justified warping. Int. J. Comput. Vis. 67(2), 141–158 (2006)

16. Sun, D., Yang, X., Liu, M.Y., Kautz, J.: PWC-Net: CNNs for optical flow using pyramid, warping, and cost volume. In: Proceedings of the IEEE Conference on Computer Vision and Pattern Recognition, pp. 8934–8943 (2018)

Progressive and Coarse-to-Fine Network for Medical Image Registration Across Phases, Modalities and Patients

Sheng Wang[1,2], Jinxin Lv[1,2], Hongkuan Shi[1,2], Yilang Wang[1,2],
Yuanhuai Liang[1,2], Zihui Ouyang[1,2], Zhiwei Wang[1,2(✉)], and Qiang Li[1,2(✉)]

[1] Britton Chance Center for Biomedical Photonics, Wuhan National
Laboratory for Optoelectronics-Huazhong University of Science and Technology,
Wuhan 430074, Hubei, China
{zwwang,liqiang8}@hust.edu.com

[2] MoE Key Laboratory for Biomedical Photonics, Collaborative Innovation
Center for Biomedical Engineering, School of Engineering Sciences,
Huazhong University of Science and Technology, Wuhan 430074, Hubei, China

Abstract. In this paper, we apply our proposed PCNet [12] on three different registration tasks assigned by the Learn2Reg challenge 2021 [1,5], i.e., CT-MR thorax-abdomen registration [3,11,14], lung inspiration-expiration registration [6] and whole brain registration [4,13], well covering three key demands in clinical practice, i.e., registration across modalities, across phases, and across patients. In these tasks, an accurate and reasonable deformation field plays a crucial role while it is often difficult to estimate in large misalignments. The core conception of our PCNet is to decompose the target deformation field into multiple sub-fields in both progressive and coarse-to-fine manners, which dramatically simplifies the direct estimation of deformation field and thus leads to a robust registration performance. The evaluation results on the three tasks demonstrate a competitive performance of PCNet and its great scalability to meet various registration demands.

Keywords: Medical image registration · Deep learning · Learn2Reg

1 Introduction

Given an image pair consisting of a fixed image and a moving image, registration requires to solve a deformation field to spatially align the fixed-moving image pair, which is a key enabling technique for varied clinical usages. The Learn2Reg challenge 2021 recently initiated three tasks representing typical clinical scenarios, i.e., CT-MR thorax-abdomen intra-patient registration (registration across modalities), CT lung inspiration-expiration registration (registration across phases) and MR whole brain registration (registration across patients). Shared by these tasks, an accurate and fast estimation of deformation field is a crucial step but challenges the academic and industrial circles.

S. Wang and·J. Lv are the co-first authors.

© Springer Nature Switzerland AG 2022
M. Aubreville et al. (Eds.): MIDOG 2021/MOOD 2021/L2R 2021, LNCS 13166, pp. 180–185, 2022.
https://doi.org/10.1007/978-3-030-97281-3_26

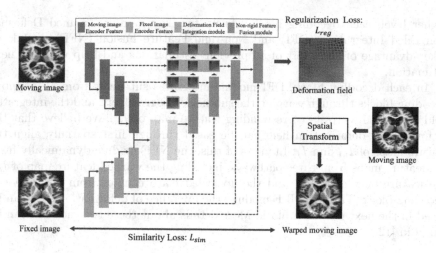

Fig. 1. The structure of proposed registration network which consist of a dual-encoder U-Net backbone, and a set of DFI and NFF modules in each decoding block.

In our recent work [12], we argued that decomposition of the target deformation field is a potential and promising solution and proposed PCNet combining the strengths of progressive registration method [15] and coarse-to-fine registration method [7]. Specifically, PCNet is built upon the backbone of dual-encoder U-net [10] and involves two key modules, i.e., deformation field integration (DFI) and non-rigid feature fusion (NFF), so as to decompose the target deformation field into several sub-fields in both progressive and coarse-to-fine manners simultaneously. By doing so, PCNet can handily predict more accurate and reasonable deformation fields with fewer parameters, and achieves competitive performance in all tasks of the Learn2Reg challenge.

2   Method

In this section, we introduce our PCNet to make this paper self-included (Sect. 2.1), and implementation details to address the three specific tasks (Sect. 2.2).

2.1   Progressive and Coarse-to-Fine Network

As shown in Fig. 1, given an unregistered pair of moving image (I_m) and fixed image (I_f), PCNet first utilizes two separate encoders to extract multi-scale features from I_m and I_f respectively. Then these multi-scale feature maps in decoding blocks is used to estimate sub-fields from coarse to fine until full-size deformation field, i.e., target deformation field are obtained. Specifically, the coarsest feature maps of I_m and I_f (gray block in Fig. 1) are combined to produce the coarsest deformation field through convolution layers. The estimation of higher level's deformation field is quite different with the lowest level. In the higher level, each decoding block sets up two key modules, named Deformation Filed Integration (DFI) and Non-rigid Feature Fusion (NFF) module to take advantage of the effort of all previous estimations and help with the next estimation.

In each decoding block, DFI module integrates sub-fields from all previous decoding blocks through warping the sub-fields progressively, and the integrated field used to warp its corresponding feature map of I_m. We believe that the estimation of deformation field can be eased through first spatially align the feature maps of I_m and I_f. In views of this, the NFF module dynamically fuses the feature maps from three pathways, including the warped feature map of I_m, the feature map of the I_f and the up-scaled feature maps from last previous decoding block. Through NFF module, the inference of deformation field can be eased in the next decoding block. More details about these two modules can be found in [12].

2.2   Implementation Details

Learn2Reg 2021 challenge consists of 3 sub-tasks: abdominal CT-MR registration (task1), inspiration-expiration lung CT registration (task2) and whole brain MR registration (task3). According to the data characteristics and task requirements, we employ different loss functions and data processing schemes for each specific sub-task.

Loss Function. For each task, the total loss (L_{total}) basically consists of image similarity loss (L_{sim}), deformation regularization loss (L_{reg}) and weakly supervised loss (L_{weakly}) if additional segmentation labels or landmarks are provided:

$$L_{total} = \alpha L_{sim} + \beta L_{reg} + \gamma L_{weakly}, \tag{1}$$

where α, β and γ represent the weight parameters of corresponding loss respectively.

The image similarity loss (L_{sim}) calculates the similarity between the warped moving image and fixed image. We use Normalization Local Correlation Coefficient (L_{LNCC}) [12] for single-modality registration as similarity loss, and Normalization Mutual Information (L_{NMI}) [4] for multi-modality registration.

The regularization loss (L_{reg}) ensures the continuity of the deformation field based on its spatial gradients [12].

If the segmentation labels or landmarks are provided for training data, we use them as weakly supervised loss (L_{weakly}) to help the training. We construct the segmentation loss (L_{seg}) based on Dice coefficient and Cross Entropy (CE) as in [9].

Target registration error (L_{TRE}) of landmarks [8] are used as an additional weakly supervise for better alignment of small structures if landmarks are provided.

We use the same regularization loss in all tasks, while similarity loss and weakly supervised loss for each task are different. Task1 is a multi-modality registration task, thus we chose L_{NMI} as the similarity loss. The weakly supervised loss is the segmentation loss L_{seg} that calculates the misalignment of correspond region in fixed and warped moving image. For task2, we use L_{LNCC} as the similarity loss. As this task focus on the registration of the internal structure in lung, we only calculate the similarity for internal part of lungs. The TRE loss L_{TRE} is utilized for better alignment of the small structure such as pulmonary vessels. We use L_{LNCC} as the similarity loss and L_{seg} for weakly supervised loss in task3. The weight parameters $\{\alpha, \beta, \gamma\}$ are set to $\{1, 1, 1\}$ in task1, $\{1, 1, 10\}$ in task2, and $\{1, 5, 1\}$ in task3.

Data Processing and Data Augmentation. The data processing methods include twice down-sampling, window adjusting and min-max normalization. Besides, we implement random rigid and elastic deformation for some tasks to augment the limited data. Table 1 shows the implementation of data processing and data augmentation for specific sub-task.

Table 1. Data processing and data augmentation for specific sub-task.

	2× Down-sampling	Window adjusting	Min-max normalization	Random rigid & elastic deformation
Task1	√	[−170, 230]	√	√
Task2	–	[0, 1100]	√	√
Task2	√	–	√	–

3 Results

Table 2 shows the result of our method for each sub-task in Learn2Reg 2021 challenge. As shown in this table, we compared the initial score with the score after registration via our method, and the results show a great improvement in all sub-tasks. According to the official final results [2], our method ranks third

Table 2. The results of Learn2Reg 2021 challenge. This table exhibits metrics including the Dice similarity coefficient of segmentation, 95% percentile of Hausdorff distance (HD) of segmentation, target registration error (TRE) of landmarks and the standard deviation of log Jacobian determinant (SDlogJ) of the deformation field. We also exhibit the time of predicting a deformation field. The last column shows the score which integrates all these metrics.

	Dice	HD	TRE	SDlogJ	Time	Initial score	Score
Task1	0.76	17.20	–	0.13	1.90	0.26	0.78
Task2	–	–	2.70	0.10	2.70	0.25	0.59
Task3	0.80	2.00	–	0.08	2.00	0.17	0.80

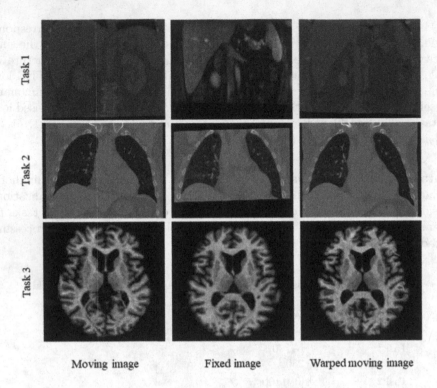

Moving image   Fixed image   Warped moving image

Fig. 2. Examples of registration results for task1, task2, and task3.

among 13 unofficial participating teams. Comparing with other top performance teams, our method obtains deformation field in relatively less time for all tasks. Figure 2 shows some visualization results of our method in three tasks. As shown in the figure, the warped moving image is able to match the fixed image with high accuracy in all tasks. These results indicate that our method works robustly in multiple situations, such as multi-modality registration, large misalignment registration, tiny and intricate structure registration.

4   Conclusion

In this paper, we introduce our learning-based registration network and evaluate its effectiveness on Learn2Reg 2021 challenge. Our method decomposes the target deformation field in both progressive and coarse-to-fine manners simultaneously, which enables the network accomplish accurate spatial alignment even when the displacement is quite large such as in the task2, or when the structure is complex such as in task3. The results of Learn2Reg 2021 challenge confirm that our method can achieve relatively satisfactory results in a variety of scenarios.

References

1. MICCAI registration challenge. https://learn2reg.grand-challenge.org/
2. Workshop-grand challenge. https://learn2reg.grand-challenge.org/Workshop/
3. Clark, K., et al.: The cancer imaging archive (TCIA): maintaining and operating a public information repository. J. Digit. Imaging **26**(6), 1045–1057 (2013)
4. Hoopes, A., Hoffmann, M., Fischl, B., Guttag, J., Dalca, A.V.: HyperMorph: amortized hyperparameter learning for image registration. In: Feragen, A., Sommer, S., Schnabel, J., Nielsen, M. (eds.) IPMI 2021. LNCS, vol. 12729, pp. 3–17. Springer, Cham (2021). https://doi.org/10.1007/978-3-030-78191-0_1
5. Hering, A., et al.: Learn2Reg: comprehensive multi-task medical image registration challenge, dataset and evaluation in the era of deep learning. arXiv preprint arXiv:2112.04489 (2021)
6. Hering, A., Murphy, K., van Ginneken, B.: Learn2Reg challenge: CT lung registration - training data, May 2020. https://doi.org/10.5281/zenodo.3835682
7. Hu, X., Kang, M., Huang, W., Scott, M.R., Wiest, R., Reyes, M.: Dual-stream pyramid registration network. In: Shen, D., et al. (eds.) MICCAI 2019. LNCS, vol. 11765, pp. 382–390. Springer, Cham (2019). https://doi.org/10.1007/978-3-030-32245-8_43
8. Hu, X., Yang, J., Yang, J.: A CNN-based approach for lung 3D-CT registration. IEEE Access **8**, 192835–192843 (2020). https://doi.org/10.1109/ACCESS.2020.3032612
9. Isensee, F., et al.: NNU-Net: self-adapting framework for u-net-based medical image segmentation. arXiv preprint arXiv:1809.10486 (2018)
10. Jaderberg, M., Simonyan, K., Zisserman, A., et al.: Spatial transformer networks. Adv. Neural. Inf. Process. Syst. **28**, 2017–2025 (2015)
11. Kavur, A.E., Selver, M.A., Dicle, O., Barış, M., Gezer, N.S.: CHAOS - combined (CT-MR) healthy abdominal organ segmentation challenge data, April 2019. https://doi.org/10.5281/zenodo.3362844
12. Lv, J., et al.: Joint progressive and coarse-to-fine registration of brain MRI via deformation field integration and non-rigid feature fusion. arXiv preprint arXiv:2109.12384 (2021)
13. Marcus, D.S., Wang, T.H., Parker, J., Csernansky, J.G., Morris, J.C., Buckner, R.L.: Open access series of imaging studies (OASIS): cross-sectional MRI data in young, middle aged, nondemented, and demented older adults. J. Cogn. Neurosci. **19**(9), 1498–1507 (2007)
14. Xu, Z., et al.: Evaluation of six registration methods for the human abdomen on clinically acquired CT. IEEE Trans. Biomed. Eng. **63**(8), 1563–1572 (2016)
15. Zhao, S., Lau, T., Luo, J., Eric, I., Chang, C., Xu, Y.: Unsupervised 3D end-to-end medical image registration with volume Tweening network. IEEE J. Biomed. Health Inform. **24**(5), 1394–1404 (2019)

Semi-supervised Multilevel Symmetric Image Registration Method for Magnetic Resonance Whole Brain Images

Marek Wodzinski[1,2]([✉]) [iD]

[1] Department of Measurement and Electronics,
AGH University of Science and Technology, Krakow, Poland
`wodzinski@agh.edu.pl`
[2] University of Applied Sciences Western Switzerland (HES-SO Valais),
Information Systems Institute, Sierre, Switzerland

Abstract. This paper describes a contribution to the second edition of the Learn2Reg challenge organized jointly with the MICCAI 2021 conference, more specifically, to the OASIS MRI task that is related to the registration of whole brain magnetic resonance images. The proposed algorithm is a multi-level, learning-based, and semi-supervised procedure. The algorithm consists of a multi-level input/output U-Net-like architecture trained with additional symmetry constraints. The method was ranked as the third-best for the brain registration task in terms of the combined challenge evaluation criteria.

Keywords: Image registration · Deep learning · Medical imaging · L2R · Learn2Reg

1 Introduction

This paper presents a contribution to the Learn2Reg challenge organized jointly with the MICCAI 2021 conference. The goal of the method is to register whole brain magnetic resonance (MR) images and to improve the alignment of small structures of variable shape and size with high precision [1]. The proposed method is a learning-based, multi-level, and semi-supervised procedure. It combines the unsupervised approach based on the MIND-loss [2], the weak supervision guided by the segmentation masks, and the symmetry enforcement based on the inverse consistency. The paper presents a slightly improved version of the method used during the previous edition of the challenge [3].

2 Methods

2.1 Method

The proposed method is a learning-based registration procedure. It consists of a multi-level U-Net-based architecture [4] with residual connections. The procedure starts with creating the resolution pyramids for both the source and the

© Springer Nature Switzerland AG 2022
M. Aubreville et al. (Eds.): MIDOG 2021/MOOD 2021/L2R 2021, LNCS 13166, pp. 186–191, 2022.
https://doi.org/10.1007/978-3-030-97281-3_27

target image. The images are then concatenated and passed to the deep network two times. First, the displacement field is calculated from the source to the target. Second, the order is reversed, and the displacement field is calculated from the target to the source. The calculated displacement fields are used to warp the source/target images and the corresponding segmentation masks. The processing pipeline and the network architecture are shown in Fig. 1. The proposed approach, apart from the inverse consistency, is based on a method dedicated to breast tumor bed localization [5].

The objective function consists of a weighted sum of the modality independent neighbourhood description self-similarity context (MIND-SSC) [2,6], the diffusion regularization, the Dice loss between the segmentation masks, and the inverse consistency of the displacement fields:

$$
\begin{aligned}
C(S, T, S_m, T_m, u_{ST}, u_{TS}) =& \frac{MIND_{SSC}(S \circ u_{ST}, T) + MIND_{SSC}(T \circ u_{TS}, S)}{2} \\
&+ \alpha \frac{R(u_{ST}) + R(u_{TS})}{2} \\
&+ \beta \frac{DSC(S_m \circ u_{ST}, T_m) + DSC(T_m \circ u_{TS}, S_m)}{2} \\
&+ \theta IC(u_{ST}, u_{TS}),
\end{aligned}
\tag{1}
$$

where S, T, S_m, T_m denotes the source, target, source mask, and target mask respectively, u_{IJ} is the dense displacement field from I to J, R denotes the diffusive regularization, IC is the inverse consistency, α, β, θ are the parameters controlling the transformation smoothness, the influence of segmentation masks, and the inverse consistency respectively. The segmentation masks are warped using the linear interpolation to make the cost function differentiable with respect to the transformation grid. The cost function is calculated for each decoder level and averaged before the backpropagation.

During the inference, the images are passed to the deep network only once to calculate the displacement field from the source to the target, without using the segmentation masks or quantitatively checking the inverse consistency.

2.2   Dataset and Experimental Setup

The dataset consists of MR images prepared for the HyperMorph article [7,8] released under the OASIS Data Use Agreement. The images show whole brains together with segmentations of 35 structures. The images are available both in an original and the preprocessed format. In this work, the preprocessed cases are used, including the skull stripping. There are 394 volumes in the training set, 19 volumes in the validation set, and 39 volumes in the test set. The training set pairs are created by matching all training cases with each other. The segmentation masks are unavailable for the test cases. A more detailed description of the dataset can be found in [1,7,8].

Training Pipeline

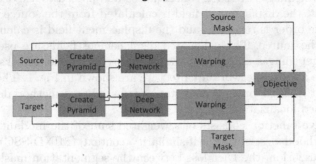

Network Architecture

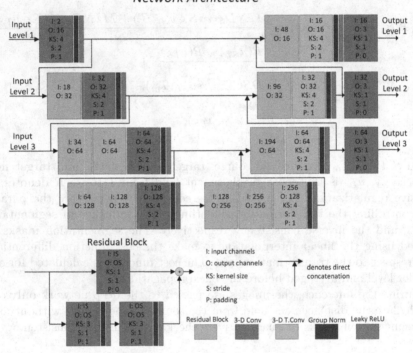

Fig. 1. Visualization of the training pipeline and the network architecture. The segmentation masks are used only for training and during the inference the images are passed to the network only once.

The training was performed for a predefined number of iterations (300), with the number of cases per iteration equal to 500, an exponentially decaying learning rate (initial learning rate: 0.001, decaying rate: 0.99), and batch size equal to 1. The network consists of 11,616,665 trainable parameters. The values for α, β, θ were 12500, 0.2 and 1500 respectively. The MIND-SSC radius and dilation were set to 2. The objective function was calculated for each encoder/decoder level separately and then summed up. The method was implemented using PyTorch [9].

3   Results

The method was evaluated using the following metrics: (i) the Dice similarity coefficient of segmentation masks (DSC), (ii) the DSC-based robustness score (DSC30 - 30% cases with the lowest DSC), (iii) the 95% percentile of Hausdorff distance (HD) of segmentations (HD95), and (iv) the standard deviation of log Jacobian determinant (SDlogJ).

The quantitative results are presented in the Table 1, together with a comparison to other participants. The table presents results for both the validation and the test set. An exemplary visualization of the registered images is shown in Fig. 2. The method was ranked as the third-best in terms of the combined challenge evaluation criteria for the registration of whole brain MR images. A more detailed comparison is available on the challenge website [1], and will be summarized in the challenge overview article [10].

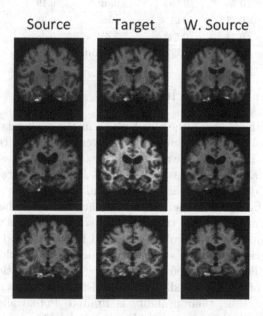

Source   Target   W. Source

Fig. 2. Exemplary visualization of the registration results for three test cases. The columns (from left) show the middle slice of source, target, and the registered source respectively.

Table 1. Quantitative results on the validation and test sets for the proposed method, as well as for the other participants' methods. The HD is presented in millimeters. Details about the rank calculation and the participants details are available in the challenge overview article and the challenge website [1,10].

Dataset	Avg. DSC	Avg. DSC30	Avg. HD95	Avg. SDlogJ	Time [s]	Rank
Validation (proposed)	0.83	0.81	1.71	0.05	–	–
Test (proposed)	0.79	0.63	inf	0.05	2.6	0.817
Test (proposed*)	0.79	0.63	1.8	0.05	2.6	–
cwmokab	0.82	0.68	nan	0.07	1.2	**0.943**
mattiaspaul	0.80	0.64	2.0	0.05	12.6	0.817
Driver	0.80	0.64	2.0	0.08	2.0	0.801
3idiots	0.80	0.65	2.0	0.08	1.5	0.779
PIMed	0.78	0.60	nan	0.06	14.9	0.701
Winter	0.77	0.59	2.4	0.08	2.6	0.583
MEVIS	0.77	0.59	2.3	0.07	10.4	0.564
lassehansen	0.77	0.59	2.2	0.09	23.8	0.503
Bailiang	0.67	0.44	2.9	0.04	1.4	0.449
smajjk	0.78	0.61	2.1	0.56	–	0.435
Imperial	0.76	0.59	nan	0.19	2611	0.413
AlexThorley	0.77	0.61	nan	0.31	–	0.405
vjaouen	0.74	0.54	2.5	0.08	–	0.358

* Assuming that the NaN/Inf HD95 equals to 10mm.

4   Discussion and Conclusion

The proposed method scored the third place in the task related to the registration of the whole brain MR images. The average method run-time could be further reduced by initializing the network prior to accessing the image pairs. The proposed method achieves low average SDlogJ and a small fraction of foldings, showing that the deformations are smooth and regular. The method could be further improved by a proper augmentation. The results for the test set are considerably worse compared to the training/validation sets. This shows the limited generalizability and potential problems related to the method usage on images acquired with previously unseen scanners/protocols. This could be alleviated by a proper domain adaptation before performing the registration. Nevertheless, the latent spaces of networks dedicated to registration of given organs are usually organ-specific, thus, their usability is limited. Potentially more universal approach, e.g. described in the SynthMorph article [11] could result in a more scalable and universal registration network.

All the participants methods suffer from the relatively low robustness. The DSC30 is significantly lower than the average DSC for all the proposed methods. The DSC score for majority of the methods vary between 0.78 to 0.82 while the differences between the DSC30 are way more significant. Therefore, the best-performing method (cwmokab) won due to the highest robustness and ability to handle difficult cases. Interestingly, the instance optimization-based methods achieved comparable results and the registration time at the level of several seconds making it arguable whether learning-based methods are the best approach for tasks without hard real-time requirements.

To conclude, the proposed algorithm is a multi-level, semi-supervised, learning-based registration method with additional symmetry constraints. Nevertheless, the method could still be further improved in terms of the generalizbility to new, unseen cases.

Acknowledgments. This work was funded by NCN Preludium project no. UMO-2018/29/N/ST6/00143.

References

1. Dalca, A., Hering, A., Hansen, L., Heinrich, M.: The challenge website. https://learn2reg.grand-challenge.org/. Accessed 28 Oct 2021
2. Heinrich, M., et al.: MIND: Modality independent neighbourhood descriptor for multi-modal deformable registration. Med. Image Anal. **16**(7), 1423–1435 (2012)
3. Wodzinski, M.: Multi-step, learning-based, semi-supervised image registration algorithm. In: Shusharina, N., Heinrich, M.P., Huang, R. (eds.) MICCAI 2020. LNCS, vol. 12587, pp. 94–99. Springer, Cham (2021). https://doi.org/10.1007/978-3-030-71827-5_12
4. Ronneberger, O., Fischer, P., Brox, T.: U-Net: convolutional networks for biomedical image segmentation. In: MICCAI 2015, pp. 234–241 (2015)
5. Wodzinski, M., Ciepiela, I., Kedzierawski, P., Kuszewski, T., Skalski, A.: Semi-supervised deep learning-based image registration method with volume penalty for real-time breast tumor bed localization. Sensors **21**(12), 1–14 (2021)
6. Heinrich, M.P., Hansen, L.: Highly accurate and memory efficient unsupervised learning-based discrete CT registration using 2.5D displacement search. In: Martel, A.L., et al. (eds.) MICCAI 2020. LNCS, vol. 12263, pp. 190–200. Springer, Cham (2020). https://doi.org/10.1007/978-3-030-59716-0_19
7. Marcus, D., Wang, T., Parker, J., Csernansky, J., Morris, J., Buckner, R.: Open access series of imaging studies (OASIS): cross-sectional MRI data in young, middle aged, nondemented, and demented older adults. J. Cogn. Neurosci. **19**, 1498–1507 (2007)
8. Hoopes, A., Hoffmann, M., Fischl, B., Guttag, J., Dalca, A.V.: HyperMorph: amortized hyperparameter learning for image registration. In: Feragen, A., Sommer, S., Schnabel, J., Nielsen, M. (eds.) IPMI 2021. LNCS, vol. 12729, pp. 3–17. Springer, Cham (2021). https://doi.org/10.1007/978-3-030-78191-0_1
9. Paszke, A., et al.: Pytorch: an imperative style, high-performance deep learning library. In: Wallach, H., Larochelle, H., Beygelzimer, A., d'Alché-Buc, F., Fox, E., Garnett, R. (eds.) Advances in Neural Information Processing Systems 32. Curran Associates, Inc., pp. 8024–8035 (2019)
10. Hering, A., et al.: Learn2reg: comprehensive multi-task medical image registration challenge, dataset and evaluation in the era of deep learning (2021)
11. Hoffmann, M., Billot, B., Greve, D., Iglesias, J., Fischl, B., Dalca, A.: SynthMorph: learning contrast-invariant registration without acquired images. IEEE Trans. Med. Imaging (2021). Early Access

Author Index

Printed in the United States
by Baker & Taylor Publisher Services